NUMERICAL ANALYSIS

PROCEEDINGS OF SYMPOSIA
IN APPLIED MATHEMATICS
Volume XXII

NUMERICAL ANALYSIS

AMERICAN MATHEMATICAL SOCIETY
PROVIDENCE, RHODE ISLAND
1978

LECTURE NOTES PREPARED FOR THE

AMERICAN MATHEMATICAL SOCIETY SHORT COURSE

NUMERICAL ANALYSIS

HELD IN ATLANTA, GEORGIA

JANUARY 3–4, 1978

EDITED BY

GENE H. GOLUB

JOSEPH OLIGER

The AMS Short Course Series is sponsored by the Society's Committee on Employment and Educational Policy (CEEP). The Series is under the direction of the Short Course Advisory Subcommittee of CEEP.

Library of Congress Cataloging in Publication Data

American Mathematical Society Short Course on Numerical Analysis, Atlanta, 1978.
Numerical analysis.

(Proceedings of symposia in applied mathematics; v. 22)
"Lecture notes prepared for the American Mathematical Society Short Course [on] Numerical Analysis, held in Atlanta, Georgia, January 3–4, 1978."
1. Numerical analysis–Addresses, essays, lectures. I. Golub, Gene Howard, 1932– II. Oliger, Joseph, 1941– III. Title. IV. Series.
QA297.A56 1978 519.4 78-11096 ISBN 0-8218-0122-8

AMS (MOS) subject classification 65-02.

Reprinted with corrections, 1980

Printed in the United States of America.

CONTENTS

Preface vii

Three research problems in numerical linear algebra
by CLEVE B. MOLER 1

A brief introduction to quasi-Newton methods
by J. E. DENNIS 19

The approximation of functions and linear functionals: Best vs. good approximation
by CARL DE BOOR 53

Numerical methods for the solution of ordinary differential equations
by JAMES M. VARAH 71

Methods for time dependent partial differential equations
by JOSEPH E. OLIGER 87

Variational methods for elliptic boundary value problems
by GEORGE J. FIX 109

Preface

This volume contains lecture notes prepared by the speakers for the American Mathematical Society Short Course on Numerical Analysis given in Atlanta, Georgia, 3-4 January 1978.

We were very pleased that the Short Course Advisory Subcommittee decided to hold a short course on Numerical Analysis, and even more pleased by the large attendance. We are indebted to our colleagues for their enthusiastic cooperation and efforts which made the Short Course and these published proceedings possible.

The choice of topics was influenced rather strongly by the subcommittee's objectives for these short courses. These objectives are that the short courses provide an entree to an area and lead up to current research problems in that area. We have tried to emphasize those areas where research activity is greatest and the present state of understanding is not satisfactory. Consequently, many classical problem areas are hardly mentioned, or ignored.

The term numerical analysis is too narrow as a description of the area as it is generally viewed today, since the construction of algorithms is an important aspect of the subject as well. In understanding why a particular algorithm "works" or "does not work" one often is led to better algorithms. Thus, the constructive and analytical aspects are not independent.

It is often said that the "computations are ahead of the analysis." This means that known algorithms perform in an inexplicable manner when they are tested on problems with known solutions. They often work better than we can guarantee them to work--our error estimates are not sharp enough or don't exist. This illustrates the strong influence that computations performed in the physical sciences and engineering have on the subject.

The papers given here are mainly of a mathematical nature. The results presented describe properties of computational methods that are only relevant in the context of that computation. It is the need to perform the computation which presents the problems to the subject and justifies it. For example, in the emerging field of Computational Physics methods are developed as they are <u>needed</u> for various problems. These methods are usually constructed via physical reasoning, experience, and intuition. They are often tested on problems with known solutions, but their validity is often judged on their behavior in physical terms. It is then the numerical analyst who attempts to give error estimates and describe the numerical behavior of these methods. The convergence results needed here differ from those of classical constructive analysis. Error estimates which hold for finite values of the discretization parameters are what are really needed, as opposed to asymptotic estimates as these parameters tend to zero. The effect of rounding errors is a central issue in numerical analysis and is a unique aspect of the subject. Algorithms which are otherwise exact may be useless because of rounding errors.

Though applications are discussed here, the important relationships between the problems, the algorithms, and the machines used for the computations which are vital to the spirit of the field cannot be found here. Numerical analysis is not a textbook subject; computational experience is essential.

We hope that these manuscripts and their bibliographies will prove useful to those who wish to learn something of the nature of numerical analysis and what some of the current problems of interest are.

Gene H. Golub
Joseph Oliger

Stanford University
July 1978

Proceedings of Symposia in Applied Mathematics
Volume 22
1978

THREE RESEARCH PROBLEMS IN NUMERICAL LINEAR ALGEBRA*

Cleve Moler

ABSTRACT. This article is intended to introduce its readers to a small sample of current research work in numerical linear algebra by describing three unsolved research problems. All three problems have arisen in the development and application of LINPACK and EISPACK, two collections of Fortran subroutines for matrix computation. Briefly, the problems involve

- Estimating the "nearness to singularity" of a matrix.
- Convergence of a method for the nonsymmetric eigenvalue problem.
- Use of matrix factorizations to approximate matrices.

1. THE LINPACK CONDITION ESTIMATOR

LINPACK is a collection of Fortran subroutines for solving various types of simultaneous linear equations and for analyzing certain types of matrices which is currently under development at Argonne National Laboratory and three universities. The simplest problem addressed by LINPACK is the solution of the equation

$$Ax = b$$

where A is a given n by n matrix of real or complex numbers and b is a given n vector. The order n is limited by the amount of memory available on any particular computer; today this usually means n is at most a few hundred.

A mathematician facing this apparently simple problem would first be concerned with existence and uniqueness of the solution x and hence be concerned about whether or not A is singular. In numerical work with inexact data and imprecise arithmetic, it is usually inappropriate to ask whether or not a matrix is singular; the distinction between singular and nonsingular becomes a bit fuzzy. It is more appropriate to formulate a quantitative notion of "nearness to singularity" and develop methods for computing it.

*Derived from a lecture given at the American Mathematical Society short course in Numerical Analysis, Atlanta, January 1978. Supported in part by NSF Grant NCS76-03052.

A closely related problem involves measuring the effects of the errors in the data and in the machine arithmetic. For simplicity, let us assume that b is known exactly, but that the elements of A may be in error, so that the system actually given to the subroutine is

$$(A+E)y = b .$$

Let $e = x-y$ be the error in the solution x resulting from the error E in the coefficient matrix. We measure the size of vectors by the ℓ_1 norm,

$$\|x\| = \sum_i |x_i|$$

because the subordinate operator norm on matrices,

$$\|A\| = \sup_{x \neq 0} \frac{\|Ax\|}{\|x\|} = \max_j \sum_i |a_{ij}|$$

is inexpensive to compute.

We will be concerned with the <u>relative</u> error in the data and solution. In order to have the error bound assume its simplest form, it is convenient to measure the error in the data relative to the theoretical coefficient matrix, that is $\|E\|/\|A\|$, but to measure the relative error in the solution relative to the perturbed solution, $\|e\|/\|y\|$. It is then an easy exercise to obtain

$$\frac{\|e\|}{\|y\|} \leq \|A\| \|A^{-1}\| \frac{\|E\|}{\|A\|} .$$

The quantity

$$\kappa(A) = \|A\| \|A^{-1}\|$$

is known as the <u>condition number</u> of A . It is the modulus of continuity of the relative error in the solution considered as a function of the relative error in the matrix. Similar error bounds involving $\kappa(A)$ can be obtained if the error is measured relative to x , or if perturbations in b are also considered.

A fundamental result of the "inverse error analysis" of J. H. Wilkinson is that the effect on the computed solution of roundoff errors in a properly designed subroutine for solving linear systems can also be regarded as resulting from some small additional error in the input matrix. Unless A is known exactly, say its elements are all integers which can be exactly represented in the computer, it is almost always true that the error in x resulting from the formation of A is more significant than the error resulting from solving the system.

As a rough rule of thumb, if we are working on a computer whose arithmetic carries, say, p significant digits, then the computed solution to a system with a condition number $\kappa(A) = 10^q$ may have only $p-q$ accurate significant digits. We must say "rule of thumb" and "may" because the error bound is only a bound and because different scalings of the matrix may alter what we mean by "significant figures." But, in general, we should become increasingly concerned as q increases toward p, and alarmed whenever q exceeds p.

Notice that we do not need to compute $\kappa(A)$ itself very accurately; we are usually only interested in its order of magnitude.

One method for solving $Ax = b$ is to compute A^{-1} and then compute $A^{-1}b$. This provides a handy formula for the solution: $x = A^{-1}b$. But consider the 1 by 1 case. Solve the equation $7x = 21$. One does <u>not</u> solve this computing $7^{-1} \cdot 21$. This would be inefficient and, in finite precision arithmetic, less accurate, than simply dividing: $x = 21/7$. Similar considerations apply to systems larger than 1 by 1. It is always more efficient and usually more accurate to solve $Ax = b$ directly with an elimination algorithm than it is to compute A^{-1} and then $A^{-1}b$. (I like to emphasize this by using the notation $A\backslash b$ and b/A for $A^{-1}b$ and bA^{-1}, respectively, but I won't pursue that point here.)

Since $\|A\|$ is readily computed, it would be possible to compute $\kappa(A)$ by finding A^{-1} and then $\|A^{-1}\|$. But, depending upon how A^{-1} is found, this will cost about 3 or 4 times as much as solving for x. Consequently, we seek a technique for obtaining a fairly accurate estimate of $\kappa(A)$ which does not require much more computer time or storage than simply solving $Ax = b$. Then, when the users of our subroutines ask us to indicate whether or not their matrices are singular, we can reply with a useful measure of nearness to singularity. The LINPACK condition estimator is a technique for doing this developed by Cline, Moler, Stewart and Wilkinson and described in a forthcoming paper.

The expensive part of $\kappa(A)$ is $\|A^{-1}\|$, which, by definition, is

$$\|A^{-1}\| = \sup_{y \neq 0} \frac{\|A^{-1}y\|}{\|y\|} .$$

The idea behind the condition estimator is to generate a particular vector y, solve $Az = y$, and then estimate

$$\|A^{-1}\| \approx \frac{\|z\|}{\|y\|} .$$

The system $Az = y$ can be solved inexpensively using whatever factorization of A is being used to solve $Ax = b$. LINPACK employs different factorizations for different types of matrices -- positive definite, symmetric indefinite, general, banded, etc. The condition estimators associated with the different factorizations differ only in programming details.

Of course, we can also write

$$\|A^{-1}\| = \sup_{z} \frac{\|z\|}{\|Az\|} .$$

In the extreme case when A is singular, there exists a vector z for which $\|z\| \neq 0$, but $\|Az\| = 0$, and $\|A^{-1}\|$ is regarded as infinite. Thus estimating condition is closely related to finding null vectors of singular matrices. If A is not singular, our goal is to find an "approximate null vector" so that

$\|z\|/\|Az\|$ is as large as possible.

The traditional method for solving singular, homogeneous systems

$$Az = 0 \ , \quad z \neq 0$$

is to pick an index k, set the k-th component z_k to a nonzero value, say -1, pick an equation number ℓ, ignore the ℓ-th equation, and solve the resulting $n-1$ by $n-1$ system. For example, if

$$A = \begin{pmatrix} 1 & 4 & 7 \\ 2 & 5 & 8 \\ 3 & 6 & 9 \end{pmatrix}$$

we might choose $k = 3$ and $\ell = 3$. Then the equation $Az = 0$ becomes

$$\left(\begin{array}{cc|c} 1 & 4 & 7 \\ 2 & 5 & 8 \\ \hline 3 & 6 & 9 \end{array}\right) \begin{pmatrix} z_1 \\ z_2 \\ \hline -1 \end{pmatrix} = \begin{pmatrix} 0 \\ 0 \\ \hline 0 \end{pmatrix}$$

$$\begin{pmatrix} 1 & 4 \\ 2 & 5 \end{pmatrix} \begin{pmatrix} z_1 \\ z_2 \end{pmatrix} = \begin{pmatrix} 7 \\ 8 \end{pmatrix}$$

$$z = \begin{pmatrix} -1 \\ 2 \\ -1 \end{pmatrix} .$$

The practical difficulties of this approach include: what k, what ℓ, and what if the $n-1$ by $n-1$ system is also singular?

A technique which is surprisingly more effective in practice is known as "inverse iteration," although this is a misnomer in our case because no actual iteration is involved.

In outline, inverse iteration goes as follows:

1. Pretend A is nonsingular.
2. Pick a random vector y.
3. Solve $Az = y$.
4. Theoretically, some components of z will usually be infinite; computationally, $\|z\|$ will usually be very large.
5. Normalize z and return $\frac{z}{\|z\|}$ as the desired solution.

Since

$$A\left(\frac{z}{\|z\|}\right) = \frac{y}{\|z\|} \approx 0 \ ,$$

the normalized z will usually be an excellent approximation to a null vector. Let us illustrate this on our example with

$$A = \begin{pmatrix} 1 & 4 & 7 \\ 2 & 5 & 8 \\ 3 & 6 & 9 \end{pmatrix} .$$

We will choose y to be the "random" vector

$$y = \begin{pmatrix} e \\ 0 \\ \pi \end{pmatrix}$$

and carry out our computations on a hypothetical computer which uses roughly three significant digits in its "floating point" or scientific notation arithmetic. In solving $Az = y$, we first "pivot" or interchange the equations so that the element 3 in A is moved to the upper left hand corner. The augmented matrix is

$$\begin{pmatrix} 3 & 6 & 9 & 3.14 \\ 2 & 5 & 8 & 0.00 \\ 1 & 4 & 7 & 2.72 \end{pmatrix} .$$

The elimination, or reduction to a row echelon form (not the reduced row echelon form -- that's more work than necessary), produces

$$\begin{pmatrix} 3 & 6 & 9 & 3.14 \\ 0 & 2.002 & 4.003 & 1.67 \\ 0 & 0 & .5\cdot 10^{-3} & -2.92 \end{pmatrix} .$$

Since we did not do the arithmetic exactly, the 3,3 element of the reduced augmented matrix is small but not zero. (If we accidentally produced an exact zero, we might replace it by any small number.) Unless we made a very special choice for y, the 3,4 element of the reduced augmented matrix will not be small. The key step is now the division

$$z_3 = \frac{-2.92}{.5\cdot 10^{-3}} = -5.85\cdot 10^{3} .$$

This value is used in the second and then the first equation to obtain

$$z = \begin{pmatrix} -5.83\cdot 10^{3} \\ 11.69\cdot 10^{3} \\ -5.85\cdot 10^{3} \end{pmatrix} .$$

The elements of z are roughly the size of the reciprocal of the accuracy parameter of our "computer", which is exactly what we wanted. The normalized z,

$$\frac{z}{\|z\|} \sim \begin{pmatrix} 1.000 \\ -2.005 \\ 1.003 \end{pmatrix}$$

is within "machine accuracy" of being a null vector for A.

An interesting aspect of this method is that it relies on the special nature of the errors introduced by elimination methods for solving systems of equations. Theoretically, the system $Az = y$ does not usually have a solution and so the computed solution "blows up." But it blows up in exactly the right direction -- it produces a vector in the null space of A.

Of course, the trick is to choose a good vector y. Almost any vector will work pretty well, but it is possible to make a choice which is entirely deficient in the desired components. A "bad" y will be signalled by the fact that $\|z\|$ is not big enough. One might be tempted in this case to simply repeat the process, using z as a new right hand side. This repetition is the iterative part of inverse iteration, but Wilkinson has shown that it is usually better to start over again with a different vector y than to iterate.

The failure of one step of inverse iteration can be illustrated with the following example.

$$A = \begin{pmatrix} -149 & -50 & -154 \\ 537 & 180 & 546 \\ -27 & -9 & -25 \end{pmatrix}$$

The matrix is not singular; in fact its determinant is 6. But it is close to singular. This can be demonstrated by the fact that

$$\begin{pmatrix} -149.00060.. & -49.99807.. & -154.00004.. \\ 536.99980.. & 180.00063.. & 545.99998.. \\ -27.00061.. & -8.99803.. & -25.00004.. \end{pmatrix}$$

is singular. We wish to find a near-null vector for A. This is a vector z so that $\|Az\|$ is much smaller than $\|z\|$.

If we choose $y = (e,\sqrt{2},\pi)^T$, solve $Az = y$, and normalize the resulting solution so that $\|z\| = 1$, we obtain

$$z = \begin{pmatrix} 0.2348 \\ -0.7492 \\ 0.0161 \end{pmatrix}.$$

This is a satisfactory solution because $\|Az\| = 0.003885$ is about the minimum of $\|Ax\|$ over all x with $\|x\| = 1$. However, if we simply alter the sign of y_3, so that $y = (e,\sqrt{2},-\pi)^T$ and carry out the process again, we obtain

$$z = \begin{pmatrix} -0.0443 \\ 0.7518 \\ -0.2039 \end{pmatrix}.$$

This is a very bad result because $\|Az\| = 1.0949$ is over 3 orders of magnitude too large.

To get some idea of why inverse iteration may fail, and why simply iterating might not help, consider the eigenvalues λ_j and eigenvectors x_j of A. Assume A has a full set of independent eigenvectors and express the starting vector y as a linear combination of them.

$$y = \sum_j \gamma_j x_j$$

Then the solution z to $Az = y$ can be expressed as

$$z = \sum_j \frac{\gamma_j}{\lambda_j} x_j \ .$$

We hope that $\|z\|$ is much larger than $\|y\|$. This would be achieved if any of the eigenvalues were small and the corresponding coefficient were nonzero. But, for nonnormal matrices, the size of the eigenvalues has little to do with the condition or nearness to singularity. A matrix can be nearly singular, but not have any particularly small eigenvalues. Our particular matrix was contrived so that its eigenvalues are 1, 2, and 3 -- none of them are small. Moreover, iterating the process once more simply gives

$$\sum_j \frac{\gamma_j}{\lambda_j^k} v_j$$

which may not be much of an improvement.

We are being pretty vague about what we mean by small. To be more precise involves considering such things as the scaling of A and the distribution of the coefficients α_j as y varies. We will not go into detail here.

A better understanding of the source of the difficulties, and a clue to their cure, can be obtained with the <u>singular value decomposition</u> of A . Here A is expressed as a product

$$A = U\Sigma V^T$$

where U and V are n by n orthogonal matrices with columns u_j and v_j and Σ is an n by n diagonal matrix with positive diagonal elements σ_j which are the <u>singular values</u>. It is not hard to see that the σ_j are the square roots of the eigenvalues of either of the symmetric, positive semi-definite matrices AA^T or A^TA and that u_j and v_j are particular choices of the corresponding eigenvectors.

Consider the following two-step process involving both A and its transpose, A^T . Start with any vector e . First solve $A^Ty = e$. Then solve $Az = y$. If the factorization used to solve $Ax = b$ is saved, it can be used to solve both these equations at little additional cost. In fact, for an n by n matrix, the amount of work required to solve a single system of equations is proportional to n^3 , but the amount of work required to solve additional systems involving the same matrix, or its transpose, is only proportional to n^2 .

In contrast to the eigenvectors, the singular vectors u_j are orthogonal and hence linearly independent, so the starting vector e can always be expressed as a linear combination of them.

$$e = \sum_j \alpha_j u_j$$

Then

$$y = \sum_j \frac{\alpha_j}{\sigma_j} v_j$$

$$z = \sum_j \frac{\alpha_j}{\sigma_j^2} u_j$$

Now comes a key point. Any matrix which is close to singular <u>must</u> have a small singular value. In fact, in the spectral norm, the smallest singular value is the distance to the set of singular matrices. Let σ_n be the smallest singular value. The desired growth in going from y to z can be achieved by insuring that the corresponding coefficient α_n is nonzero. Since the u_j are orthogonal and e is random, the chances of α_n being too small are remote.

The LINPACK condition estimator goes one step further. The initial vector e is not chosen randomly, but rather by a fairly complicated process that involves the particular factorization of A used to solve linear systems. Different types of matrices have different factorizations and hence different choices of e's , but in all cases the intent is to achieve growth in going from e to y .

In summary, then, the condition estimator works as follows:

1. Compute $\|A\|$.
2. Factor A .
3. Use the factors to define e .
4. Use the factors to solve $A^T y = e$.
5. Use the factors to solve $Az = y$.
6. Estimate $\|A^{-1}\| \approx \frac{\|z\|}{\|y\|}$.
7. Estimate $\kappa(A) = \|A\| \cdot (\text{estimate of } \|A^{-1}\|)$.

Note that the estimate is always a <u>lower bound</u> for the true $\kappa(A)$.

For our example matrix, the singular values are

817.760, 2.47497 and 0.0029645

whereas the eigenvalues are 3, 2, and 1. The product of the singular values must also be 6, the same as the absolute value of the product of the eigenvalues, but their distribution is quite different. The three starting vectors $(e, \sqrt{2}, \pi)^T$, $(e, \sqrt{2}, -\pi)^T$, and the vector used by LINPACK all produce essentially the same estimate for $\kappa(A)$, namely $1.94 \cdot 10^5$. The actual $\kappa(A)$ is $2.18 \cdot 10^5$.

All three starting vectors also lead to essentially the same approximate null vector,

$$z = \begin{pmatrix} -0.2348 \\ 0.7492 \\ -0.0160 \end{pmatrix}$$

A two-step process that does not involve A^T -- pick e, solve $Ay = e$, solve $Az = y$ -- produces very poor estimates of $\kappa(A)$ and very poor approximate null vectors.

The LINPACK subroutines actually produce a quantity RCOND which approximates $1/\kappa(A)$ because this can be set to zero if exact singularity is detected, or if the estimate for $\kappa(A)$ is too large to be expressed in the floating point arithmetic.

As an important by-product, the subroutines also return the vector z. This is an approximate null vector in the sense that

$$\|Az\| \leq \text{RCOND} \cdot \|A\| \cdot \|z\| .$$

The description given here, as well as the somewhat more detailed one given in the paper by Cline, Moler, Stewart and Wilkinson, is certainly far from a proof that the condition estimate will always be an accurate one. But it, together with a fairly extensive set of numerical experiments, have convinced us that it is quite reliable.

That brings us to our first research problem: how well does the estimator really work and how can it be made any better without increasing the cost. We can formulate two specific subproblems.

1. Let $\mathcal{M}_n$ be the set of all n by n real or complex matrices. Let $\text{COND}(A) = 1/\text{RCOND}$ be the estimate provided by the LINPACK subroutine applied to A and let $\kappa(A)$ be the true condition number of A (in the ℓ_1 norm). What is

$$\min_{A \in \mathcal{M}_n} \frac{\text{COND}(A)}{\kappa(A)} ?$$

2. Let $\mathcal{A}_n$ be the set of algorithms $\tilde{\kappa}$ with $O(n^2)$ work beyond the factorization required to solve a linear system. What is

$$\max_{\tilde{\kappa} \in \mathcal{A}_n} \min_{A \in \mathcal{M}_n} \frac{\tilde{\kappa}(A)}{\kappa(A)} ?$$

2. CONVERGENCE OF THE QR ALGORITHM

The QR algorithm is a method for computing matrix eigenvalues developed around 1960 by J. G. F. Francis, following the earlier LR algorithm of H. Rutishauser. The Q stands for orthogonal and the R for right triangular. The algorithm is now the most widely used method for computing eigenvalues of general matrices. Several subroutines in EISPACK implement the method for different types of matrices. We are concerned here particularly with the real, nonsymmetric QR algorithm as implemented in subroutines HQR and HQR2, which are Fortrans translations of Algol procedures developed by J. H. Wilkinson and his colleagues. The input to either subroutine is any n by n real matrix which can be stored in the computer memory. HQR finds the eigenvalues of the matrix and HQR2 find its eigenvalues and eigenvectors. (If a full set of eigenvectors does not exist, multiple copies of some eigenvectors will be produced, but that's not important for the discussion here.)

Considerable experimental evidence, gathered over many years of experience with HQR and related subroutines, has shown the QR algorithm to be exceptionally effective. But the theoretical basis for the algorithm actually used in practice is still incomplete. We are in the interesting situation of having no proof that the algorithm is guaranteed to work, but having no examples (as far as I know) on which it fails.

A complex matrix U with complex conjugate transpose U^H is <u>unitary</u> if $U^H U = I$ and a real matrix U with transpose U^T is <u>orthogonal</u> if $U^T U = I$. A matrix T is <u>upper triangular</u> if $t_{ij} = 0$ for $i > j$. (All matrices in this section are square.)

The background for the QR algorithm is provided by a 1909 lemma of Schur. For any complex matrix A there exists a unitary matrix U so that $U^H AU = T$ is upper triangular. In particular, the eigenvalues of A occur on the diagonal of T.

In order to avoid complex arithmetic, which is expensive in terms of both computer time and storage, subroutines based on the QR algorithm do not actually produce a triangular matrix, but rather a "quasi-triangular" matrix with one by one and two by two blocks on its diagonal. More precisely, T is quasi-triangular if $t_{ij} = 0$ for $i > j+1$ and if $t_{i,i-1} \neq 0$ then $t_{i+1,i} = 0$ and $t_{i-1,i-2} = 0$. A 1931 paper by F. D. Murnaghan and A. Winter pointed out that for any real A there is a real orthogonal matrix U so that $U^T AU = T$ is quasi-triangular. The one by one diagonal blocks in T are real eigenvalues of A ; the eigenvalues of the two by two diagonal blocks in T are pairs of complex conjugate eigenvalues of A . The QR algorithm produces a sequence of orthogonally similar matrices which, if they converge, converge to the canonical form whose existence was established by Murhaghan and Winter.

The QR algorithm itself is usually preceeded by a finite algorithm which produces an orthogonal matrix P , a product of $n-2$ special orthogonal matrices known as Householder reflections, so that $\tilde{A} = PAP^T$ is in <u>Hessenberg</u> form, that is $\tilde{a}_{ij} = 0$ for $i > j+1$. Let $e_i = \tilde{a}_{i,i-1}$ be the 'extra' subdiagonal elements. If any $e_i = 0$, the matrix breaks into the direct sum of smaller blocks and the eigenvalues of these blocks can be computed instead. Otherwise, if $e_i \neq 0$ for all i , the matrix is said to be in <u>unreduced</u> Hessenberg form.

The task of the QR algorithm is to produce further orthogonal similarity transformations which reduce the magnitude of the subdiagonals e_i . Eventually enough of the e_i will be below some tolerance, such as the roundoff level of the particular computer being used, that the matrix can be declared to be quasi-triangular. Then the eigenvalues can easily be extracted and the corresponding eigenvectors can be obtained if desired.

(We note in passing that we have said nothing so far about computing zeros of a characteristic polynomial. The characteristic polynomial is a valuable theoretical device, but attempts to use it as the basis for a practical general purpose eigenvalue algorithm lead to inevitable difficulties involving both speed and accuracy. In fact, in some situations, it is reasonable to find the zeros of a polynomial by setting up the corresponding companion matrix and using the QR algorithm to find its eigenvalues.)

In general outline, the QR algorithm is fairly easy to describe. We present here what is known as the Francis, double shift, real, Hessenberg QR algorithm. Start with $A_1 = \tilde{A}$, a real, n by n, unreduced Hessenberg matrix. For $k = 1, 3, 5, \ldots$, compute A_{k+1} and A_{k+2} as follows. Let s_k and s_{k+1} be two 'shifts', namely the eigenvalues of the last two by two diagonal submatrix of A_k. Form $A_k - s_k I$ and factor it into the product of an orthogonal Q_k and an upper (or right) triangular R_k, that is $A_k - s_k I = Q_k R_k$. Formally, this is equivalent to applying the Gram-Schmidt process to transform the columns of $A_k - s_k I$ into the orthonormal columns of Q_k using linear combinations specified by the elements of R_k, but the Hessenberg form of A_k means that the orthogonalization can be done with $O(n^2)$ work instead of the $O(n^3)$ work required by Gram-Schmidt on an arbitrary matrix.

Next, multiply the factors together in the opposite order and add the shift back in, that is $A_{k+1} = R_k Q_k + s_k I$. It is not hard to show that A_{k+1} is similar to A_k, and hence has the same eigenvalues as A, and that A_{k+1} is again in unreduced Hessenberg form. The process is repeated with s_{k+1} to produce A_{k+2}, that $A_{k+1} - s_{k+1} I = Q_{k+1} R_{k+1}$ and then $A_{k+2} = R_{k+1} Q_{k+1} + s_{k+1} I$. Again, A_{k+2} has the same eigenvalues as the original A and, because we are constantly forcing things towards upper triangular form, it is 'usually' closer to triangular, or at least quasi-triangular, form than A_k.

In practice, the algorithm is not carried out in exactly this way, but rather A_{k+2} is computed directly from A_k. The intermediate A_{k+1} and the two Q's and R's never appear explicitly. It turns out that A_{k+2} can be expressed using only $s_k + s_{k+1}$ and $s_k \cdot s_{k+1}$, which are always real, and so A_{k+2} is real, even though some of the other matrices would be complex. For analysis of convergence, however, this description of the algorithm is probably the most straightforward.

Let e_i^k be the subdiagonal elements of A_k. In studying convergence of the algorithm, we are only concerned with the last two subdiagonal elements. The process is said to converge if either e_n^k or $e_{n-1}^k \to 0$ as $k \to \infty$.

If e_n^k becomes less than some tolerance, then $a_{n,n}^k$ is taken as an eigenvalue of A. If e_{n-1}^k becomes less than the tolerance, then the (possibly complex) eigenvalues of the last two by two diagonal submatrix are taken as two eigenvalues of A. In either case, the process is then started over again on the remaining submatrix of order $n-1$ or $n-2$.

A simplified, and hence somewhat impractical, form of the algorithm is well understood. This is the 'no-shift' version in which s_k and s_{k+1} are taken to be zero for all k. In 1968 B. N. Parlett established the following theorem. The QR algorithm with no shift, applied to an unreduced Hessenberg matrix, converges if and only if among each set of eigenvalues of equal magnitude, there are at most two of even and two of odd multiplicity. For example, the no-shift algorithm will converge when applied to a 6 by 6 matrix with eigenvalues (counting multiplicities) 1, 1, -1, -1, i, -i , but it will not converge when applied to a 4 by 4 matrix with eigenvalues 1, -1, i, -i . However, even when convergence occurs, the lack of shifts causes the rate of convergence to be too slow to be practical.

The double shift algorithm, as we have described it so far, can get into trouble, too. For example, let

$$A = A_1 = \begin{pmatrix} 0 & 0 & 1 \\ 1 & 0 & 0 \\ 0 & 1 & 0 \end{pmatrix}$$

Then s_1 and s_2, which are the eigenvalues of $\begin{pmatrix} 0 & 0 \\ 1 & 0 \end{pmatrix}$, are both zero. Moreover, $Q_1 = Q_2 = A$, $R_1 = R_2 = I$ and so $A_2 = A_3 = A$. In other words, nothing happens. This is consistent with Parlett's theorem because the shifts are zero and A has three simple eigenvalues of equal magnitude, so there can be no convergence.

Wilkinson was certainly aware of this example when he wrote HQR, so he included a special 'ad hoc shift' to combat it. At the steps when the iteration counter, k, is 10 or 20, the shifts are not computed from the lower two by two, but rather taken so that

$$s_k \cdot s_{k+1} = (|e_n^k| + |e_{n-1}^k|)^2$$

$$s_k + s_{k+1} = 1.5(|e_n^k| + |e_{n-1}^k|)$$

The precise form of the ad hoc shift should not be regarded too seriously. It is simply a device for shifting the spectrum by some fairly arbitrary amount and giving the basic algorithm another chance from a different starting point.

This, then, is the situation today, after over 15 years of experience and study. The unsymmetric, Hessenberg QR algorithm with no shifts is not really practical and does not converge in all situations, but necessary and sufficient conditions for convergence are known. The unsymmetric, Hessenberg QR algorithm with shifts from the lower two by two is usually quite effective, but there are a few counterexamples for which it does not converge. No necessary or sufficient conditions for convergence are known. The Wilkinson ad hoc shift, or any of several similar devices, appears experimentally to take care of all the counterexamples, but there is no supporting theory.

The research problem is: does it work or doesn't it. If it does, prove it. If it doesn't, fix it, and then prove that it works.

For symmetric matrices, by the way, the situation is quite different. A symmetric Hessenberg matrix is tridiagonal and has real eigenvalues. Both of these properties help simplify both the algorithm and the theory greatly. In 1968 Wilkinson proved that the symmetric, tridiagonal QR algorithm with single shift from the lower two by two is globally convergent.

3. LOW RANK APPROXIMATIONS TO LARGE ARRAYS

Our third problem is rather vague and ill-formed, but also quite open-ended and potentially very interesting. It concerns the 'information' content of large arrays of numbers. We say array, rather than matrix, because the individual elements are important in their own right -- we are definitely <u>not</u> thinking of representations of abstract linear transformations.

The arrays need not even be square, but we will limit ourselves to the square case here to simplify notation slightly. Let A be an n by n array of real numbers. If n is quite large, say several hundred or more, it is expensive to store A in a computer, it is expensive to transmit A from one storage medium to another, and it is difficult and expensive to extract the pertinent information in A. Consequently we consider approximating A by another matrix A_k which can be stored, transmitted and analyzed more easily, but which hopefully still contains most of the useful information in A. In several different applications, such an approximation can be obtained using the singular value decomposition of A.

Let $A = U\Sigma V^T$ be the SVD of A. Let σ_j, $j = 1,\ldots, n$ be the singular values, that is the diagonal elements of Σ, ordered in decreasing order. Let u_j and v_j be the orthonormal columns of U and V respectively. These quantities result from the 'dual eigenvalue problem'

$$Av_j = \sigma_j u_j$$

$$A^T u_j = \sigma_j v_j$$

Let E_j be the rank one matrix obtained by taking the outer product of u_j and v_j, that is

$$E_j = u_j v_j^T$$

The equation $A = U\Sigma V^T$ can then be rewritten as the 'outer product expansion'

$$A = \sigma_1 E_1 + \sigma_2 E_2 + \ldots + \sigma_n E_n$$

One important fact about the E_j's is that each one requires only $2n$ storage locations, namely n locations for u_j and n locations for v_j, rather than the n^2 locations required by a matrix of full rank. Other properties of the E_j's include:

$$\|E_j\|_F = (\sum_{i,j} e_{ij}^2)^{1/2} = 1$$

$$\|E_j\| = 1 \quad \text{(the } \ell_2 \text{ operator norm)}$$

$$E_i E_j^T = 0 \text{ if } i \neq j$$

$$E_1 E_1^T + E_2 E_2^T + \ldots + E_n E_n^T = I$$

The matrix approximation we have in mind is obtained by truncating the outer product expansion after k terms ,

$$A_k = \sigma_1 E_1 + \sigma_2 E_2 + \dots + \sigma_n E_n$$

The approximation A_k requires only $(2n+1)k$ storage locations. Moreover, it is not hard to show that, using the ℓ_2 operator norm,

$$\|A - A_k\| = \sigma_{k+1}$$

Consequently, if the σ_k decrease fairly rapidly as k increases, it will be possible to obtain a fairly accurate approximation which requires relatively little storage.

One particularly interesting application involves digital image processing. It is described in a paper by H. C. Andrews and C. L. Patterson and in a book by Andrews and B. R. Hunt. A photograph or a TV image is digitized by covering it with an n by n grid and then sampling some measure of brightness in each cell of the grid. The result is an n by n array of nonnegative numbers. The accuracy of each element is only about 1% or 0.1% and the value of any element is usually close to the values of the elements near it in the array. Thus the matrix has some fairly special properties which are difficult to characterize in matrix terms.

Typical values of n in the image processing business are powers of 2 in the range from 64 to 1024. The goal, of course, is to reduce storage and transmission time from n^2 to a modest multiple of n . Computing the SVD of a 1024 by 1024 matrix is not really practical with present computers, so some substitutes for the SVD have to be made at this end of the range. But the interesting mathematical questions occur already when n is only a few hundred. Several examples of photographs approximated by truncated outer product expansions are included in the work of Andrews and his colleagues.

It turns out that the singular values of many pictures decrease quite rapidly with increasing index, so useful approximations can be obtained which require only 1/10 or less of the original storage. How generally true is this? What properties of the image insure than σ_k decreases rapidly with increasing k ?

To give the problem a somewhat more mathematical setting, let $f(x,y)$ be a function defined, say, on the unit square. Pick an integer n, let $h = 1/n$ be the grid size and let A be the matrix with elements

$$a_{ij} = f(ih,jh) \quad , \; i,j = 1,\ldots, n$$

How does smoothness and other properties of f affect the distribution of the singular values of A ?

The accuracy of matrix approximations obtained from the SVD can be discussed most readily using the ℓ_2 norm or the F norm defined above. But these are probably <u>not</u> the correct norms for discussing the accuracy of image approximations. What norm does the human optical system employ when comparing two photographs?

Superficially at least, the outer product expansion is related to Fourier series. Notions of orthogonality are present in both and increasing the number of terms increases the accuracy. One is tempted to use terms like 'low frequency' and 'high frequency' for the E_j associated with large and small singular values, but it is not clear how far this analogy can be carried.

REFERENCES

H. C. Andrews and B. R. Hunt, Digital Image Processing, Prentice-Hall, Englewood Cliffs, 1977.

H. C. Andrews and C. L. Patterson, ''Outer product expansions and their uses in digital image processing'', Amer. Math. Monthly 82, 1-12, 1975.

A. K. Cline, C. B. Moler, G. W. Stewart and J. H. Wilkinson, ''An estimate for the condition number of a matrix'', SIAM J. Numer. Anal. 16, 368-375, 1979.

J. J. Dongarra, J. R. Bunch, C. B. Moler and G. W. Stewart, LINPACK Users' Guide, SIAM, Philadelphia, 1979.

G. E. Forsythe, M. A. Malcolm and C. B. Moler, Computer Methods for Mathematical Computations, Prentice-Hall, Englewood Cliffs, 1977.

B. S. Garbow, J. M. Boyle, J. J. Dongarra, and C. B. Moler, Matrix Eigensystem Routines -- EISPACK Guide Extension, Lecture Notes in Computer Science, vol. 51, Springer-Verlag, Heidelberg, 1977.

F. D. Murnaghan and A. Wintner, ''A canonical form for real matrices under orthogonal transformations'', Proc. N. A. S. 17, 417-420, 1931.

B. N. Parlett, ''Global convergence of the basic QR algorithm on Hessenberg matrices'', *Math. Comp.* 22, 803-818, 1968.

B. T. Smith, J. M. Boyle, J. J. Dongarra, B. S. Garbow, Y. Ikebe, V. C. Klema, and C. B. Moler, *Matrix Eigensystem Routines -- EISPACK Guide*, Second Edition, Lecture Notes in Computer Science, vol. 6, Springer-Verlag, Heidelberg, 1976.

G. W. Stewart, *Introduction to Matrix Computations*, Academic Press, New York and London, 1973.

J. H. Wilkinson, *The Algebraic Eigenvalue Problem*, Oxford Univ. Press, (Clarendon), London and New York, 1963.

J. H. Wilkinson and C. Reinsch, editors, *Handbook for Automatic Computation, volume II, Linear Algebra*, Springer-Verlag, Heidelberg, 1971.

J. H. Wilkinson, ''Global convergence of tridiagonal QR algorithm with origin shifts'', *Lin. Alg. and Appl.* 1, 409-420, 1968.

DEPARTMENT OF MATHEMATICS AND STATISTICS
UNIVERSITY OF NEW MEXICO
Albuquerque, New Mexico

Proceedings of Symposia in Applied Mathematics
Volume 22
1978

A BRIEF INTRODUCTION TO QUASI-NEWTON METHODS[1]

by

J. E. Dennis Jr.[2,3]

Cornell University
Ithaca, NY 14853

0. Introduction

This section was written last as most introductions are and so it also functions as a disclaimer for omissions, misconceptions and misrepresentations.

We try below to present a current, hurried, view of a field in flux, because of progress and changing standards. In this field theorems serve to document progress and point promising future routes. Our purpose is to solve problems and our goal is to develope better algorithms certified by realistic theorems and implemented in robust software. We are beginning to see young people coming into the field who have the breadth to do all these things but there is still room for specialized contributions.

1. The Problem

The problems considered here might be looked at by a novice to computation as being really only one problem since they are mathematically equivalent in a fairly meaningful sense. From our algorithmic point of view, there is a significant gain in exploitable structure as we pass from the problem of simultan-

---

[1]Prepared for AMS Shortcourse on Numerical Analysis, Atlanta, Jan. 1978.

[2]Research supported by NSF Grant MCS76-00324.

[3]Current address: Department of Mathematical Sciences, RICE University, Houston, Texas 77001.

eous nonlinear equations to nonlinear least squares via unconstrained minimization. We mean by this that certain features of the problem lend themselves to more efficient computer implementations of the class of algorithms outlined in section 2.

The first problem we will consider is certainly an instance of one of the most basic in mathematics.

Let Ω be an open convex set in $\mathbb{R}^n$.

NLEQ: Given $F : \Omega \subset \mathbb{R}^n \to \mathbb{R}^n$ find $x^* \in \Omega$ for which $F(x^*) = 0$.

We will use the notation T for transpose and for $x \in \mathbb{R}^n$, $x = (x^1, x^2, \ldots, x^n)^T$ and $F(x) = (f^1(x), f^2(x), \ldots, f^n(x))^T$. For derivatives, we write $F'(x) = J(x) = (\partial f^i(x)/\partial x^j)$ and for gradients $\nabla f(x) = (\partial f(x)/\partial x^1, \ldots, \partial f(x)/\partial x^n)^T$, where $f(x)$ is a real-valued function of the vector variable x. Unless otherwise specified, the Jacobian matrix $J(x)$ will be assumed to be of full rank at x^* and Lipschitz continuous in Ω.

The second problem has two additional algorithmically useful features.

UCMIN: Given $f : \Omega \to \mathbb{R}^1$, find $x^* \in \Omega$, a local minimizer of f, i.e., for some $\varepsilon > 0$, $x \in N(x^*, \varepsilon)$ implies $f(x) \geq f(x^*)$.

We will always assume f has a Lipschitz continuous second derivative and use the notation $f''(x) = \nabla^2 f(x) = (\partial^2 f(x)/\partial x^i \partial x^j)$. Notice that we will find x^*, if it exists, among the zeros of the system of nonlinear equations $\nabla f(x) = 0$. It will be useful that the Jacobian of this system is symmetric and is positive definite near a solution for most practical problems. We will assume that $\nabla^2 f(x)$, the Hessian, is symmetric and positive definite at x^*, as well as Lipschitz continuous in Ω.

Another very important point is the utility of f itself as a means of choosing between approximate solutions. In algorithms for NLEQ, it is often the case that some norm of F is used in this manner but this leads to the danger of finding a local minimizer of $||F(x)||$ rather than a zero. When the l_2 norm is used, this is an important practical problem in its own right.

NL2: Given $R : \Omega \subset \mathbb{R}^p \to \mathbb{R}^n$, $p < n$, minimize

$$\phi(x) = \frac{1}{2}R(x)^T R(x) = \frac{1}{2}\sum_1^n (r^i(x))^2.$$

This nonlinear least squares problem arises frequently in practice, especially in the context of best fitting n data points from a family of functions depending on p parameters. Furthermore, there are two quite different cases. Sometimes the final fit is very good in the sense that $R(x^*)$ is nearly zero. On the other hand, in many real problems data errors preclude small final residuals $R(x^*)$. It turns out to be useful to know which to expect since the small residual case can be handled with a much simpler algorithm if it is expected. We can indicate why this is so and establish some more notation at the same time.

It is easy to see that with $J(x) = R'(x)$, $\nabla\phi(x) = J(x)^T R(x)$ and $\nabla^2\phi(x) = J(x)^T J(x) + \sum_1^n r^i(x)\nabla^2 r^i(x)$. Thus, when $J(x)$ is computed or approximated for the gradient evaluation, it is available to form the first term in the expression given above for the Hessian. In the small residual case, this term suffices for the algorithms we will consider and the savings is considerable. The gradient also has useful structure which can be exploited in the nonzero residual case to yield a scale-free solution test. Much more fuss is made of all this in [25] where an introduction is also given to the interesting idea of exploiting linearity in some of the parameters. See [42], [45], [47], [58], [66], [75].

In practice, these three problems are sometimes posed with x restricted to lie not in an open set Ω but in some set Γ defined by inequality and equality constraints. Often this causes the significant complications which arise when x^* is on the boundary of Γ. Recently, there has been very rapid development in this area but since all that work is based on techniques for the corresponding unconstrained problem and since [34], [65] and [72] are such excellent current surveys we do not consider constrained problems here.

2. The One Basic Algorithm

The algorithms in most common usage for all these problems are in virtually all cases variations of Newton's method for NLEQ. This algorithm, also called the Newton-Raphsen method, can be written as:

$$\text{Given } x_0,\ x_{k+1} = x_k - J(x_k)^{-1}F(x_k),\ k=1,2,\ldots ,$$

Notice that we have elected to use subscripts for the iteration counter and superscripts for vector components.

The iteration formula above is certainly not the way Newton's method should be implemented on a computer. The following form will be useful for our discussion and it is much more like an actual implementation. We will call such algorithms quasi-Newton.

0) Given x_k, $F(x_k)$ and $J(x_k)$;

i) Solve the nxn linear system $J(x_k)s_k^N = -F(x_k)$ for the Newton step s_k^N.

ii) Using s_k^N and perhaps some other values of $F(x)$, choose x_{k+1}.

iii) Evaluate $F(x_{k+1})$ and test for convergence and non-convergence. Either terminate the computation or proceed to (iv).

iv) Evaluate (or approximate) $J(x_{k+1})$, set the counter to k+1 and return to (i).

Each of these four steps is still rather vague and the resolution of these vagaries has a profound impact on the performance of the implementation. The solution of the linear system in (i) is dealt with extensively elsewhere and briefly in Appendix A, so we will only make occasional remarks in context about particular advantages of certain algorithms in certain cases. For example, when $J(x_k) = \nabla^2 f(x_k)$ a symmetric indefinite factorization can be used to exploit symmetry and perhaps even a Cholesky factorization can be used when f is convex. It does seem appropriate to make some other remarks here however.

On the face of it, Newton's method requires the Jacobian matrix to be nonsingular at each iterate. When x_k is near x^*, the same concerns about well-posedness of the problem guide us in the case when F is an affine mapping and the same remedies such as discarding some combinations of the variables to make the problem acceptably well-conditioned seem reasonable. When x_k is not so near x^*, it is not unusual to encounter numerical singularity in $J(x_k)$ even for the problems where $J(x^*)$ is well-conditioned. It seems certainly wrong to discard a subset of the variables because of such temporary difficulties. How to cheaply deal with a numerically singular Jacobian in the early stages of the iteration is a very important practical problem which seems closely tied to the success of some other aspects of the algorithm such as (ii). See [25].

The choice of termination criteria is fraught with peril and again we refer to [25].

The traditional areas of research on quasi-Newton methods are steps (ii) and (iv). The reason is that for real problems, evaluations of F and J dominate the cost of solution and so it is in these steps that the potential savings is greatest.

It is worth remarking that the value of $F(x_{k+1})$ will often be available as a byproduct of the choice of x_{k+1}. However, we put this in step (iii) because $F(x_{k+1})$ must be evaluated anyway to carry out the iteration and so it should not be charged to whatever technique is used in (ii) to pick x_{k+1}. Furthermore, this makes it easier to put the unvarnished Newton method, where $x_{k+1} = x_k + s_k^N$, into our framework. It is absolutely essential that step (ii) allow a full Newton step whenever possible and that step (iv) be carried out in such a way as to enable this to be done once x_k is near x^*.

Much of the remainder of this paper will be concerned with methods for the approximation of $J(x_k)$. My experience is that

$J(x_k)$ is rarely available. Even when it is, there is a need to take and code n^2 partial derivatives. This is obviously an expensive and error prone task. We will call these algorithms quasi-Newton to allow for the use of approximate derivatives.

3. The Kantorovich Theorem

One of the most beautiful theorems in numerical analysis certifies the performance of Newton's method under very reasonable hypotheses on the problem. The version given here is slightly nonstandard, it comes from applying a more standard version to $J(x_0)^{-1}F(x)$ rather than $F(x)$. We will explain later why we prefer this version.

Theorem 3.1. Let $r > 0$, $x_0 \in R^n$, $F : N(x_0,r) \to \mathbb{R}^n$ and assume that $J(x)$ exists on $N(x_0,r)$ with $J(x_0)$ invertible. Let $\gamma,\eta \geq 0$ exist such that $||J(x_0)^{-1}[J(x) - J(y)]|| \leq \gamma||x - y||$, $||J(x_0)^{-1}F(x_0)|| \leq \eta$ and $h = \gamma\eta \leq \frac{1}{2}$. Under these hypotheses, if $r \geq r_0 = (1 - \sqrt{1-2h}\,)/\gamma$ then Newton's method starting from x_0, $x_{k+1} = x_k - J(x_k)^{-1}F(x_k)$, $k=0,1,\ldots$ exists and converges to x^*, a unique zero of F in $N(x_0,r_0)$. If $h < \frac{1}{2}$ then x^* is the only zero of F in $N(x_0,r_1)$ where $r_1 = \min(r,(1 + \sqrt{1 - 2h}\,)/\gamma)$.

Furthermore, for each k, set $t_{k+1} = t_k - (\frac{1}{2}\gamma t_k^2 - t_k + \eta)/(\gamma t_k - 1)$ with $t_0 = 0$, and then $||x_k - x^*|| \leq r_0 - t_k$.

A proof of this theorem and several similiar ones can be found in [22]. The basic idea is a lovely one. It involves showing that Newton's method for $q(t) = \frac{1}{2}\gamma t^2 - t + \eta$ starting from $t_0 = 0$ majorizes the vector iteration. The invertibility of $J(x_k)$ follows from the Banach Lemma [57]. The existence of x* follows from a completeness argument similiar to the one used in the Contractive Mapping Theorem. In fact, the connection between these two results is much deeper than one might suspect at first.

There is a tendency to think in terms of applying Theorem 3.1 to a specific F and x_0 to see if F has a zero and if Newton's method starting from x_0 will be guaranteed to find it and obey the error bounds $r_0 - t_k$. In fact, it has been usefully applied this way to integral equations [13].

This author at least, now prefers a different point-of-view. This is because it is almost certainly more difficult to determine a value for γ, the Lipschitz constant, than it is to run the program and see if the iteration converges. Furthermore, a few tests dispel the notion that the error bounds are much good in general although they are exact for the majorizing scalar iteration.

We prefer to think that the theorem identifies a very broad and reasonable class of problems for which the iteration converges. The reason we like the version given here is that the sufficient condition $\gamma \cdot \eta \leq \frac{1}{2}$ can be interpreted as relating convergence to the scale of F, or the quality of the solution estimate x_0, as measured by η and γ, a relative measure of the local nonlinearity of F. Obviously $\gamma = 0$ for an affine F and Newton's method is globally convergent in such cases. As the problem becomes more nonlinear, a better initial estimate is needed for convergence.

The error bounds are, as we have said, not very good in practice. Furthermore, they only imply r-quadratic (order 2) convergence rather than the more useful q-quadratic convergence which makes Newton's method as fast as any method ever really need be. There is an extensive discussion of these notions of convergence order in [57] and a brief discussion in terms of step (iv) of the general algorithm of Section 2 in [22].

Basically an iteration has q-order at least q if $\limsup_k e_{k+1}/e_k^q$ is finite, where $e = ||x - x^*||$ and it has r-order r if there is some sequence of error bounds $< b_k >$ with q-order r. Obviously a given q-order implies at least the same r-order. In [20] we called r-order "pseudo-order" and showed the two equivalent if and only if $\limsup_k b_k/e_k$ is finite. Still, it is probably better to have an r-order result than none at all.

It is easy to obtain a much more important error estimate: under the hypothesis of Theorem 3.1, $\frac{1}{2}||x_{k+1} - x_k|| \leq e_k \leq 2||x_{k+1} - x_k||$. See [22] for a numerical comparison with $b_k = r_0 - t_k$.

4. Least Change Secant Methods

In Section 2 we mentioned the necessity for finding good cheap ways to approximate $J(x_k)$. For n = 1, the secant method for NLEQ is a very satisfactory way to do this. This classical quasi-Newton method uses $[F(x_{k+1}) - F(x_k)]/[x_{k+1} - x_k] \equiv y_k/s_k$ as its approximation to $J(x_{k+1})$. There is a beautiful economy here since $F(x_{k+1})$ is needed anyway and so no additional function or derivative work is done in step (iv).

4.1. Jacobian Approximations

In higher dimensions there is an affine set $Q(y_k, s_k)$ of nxn matrices that send s_k to y_k and thus generalize the secant approximation to $J(x_{k+1})$. A great deal of theoretical work has been done [57] on extensions of the secant method to n-dimensions by resolving this ambiguity in such a way as to achieve r-order $\frac{1}{2}(1 + 5^{\frac{1}{2}})$, the q-order of the scalar method. None of these methods has so far proven to be of much use in actually solving particular instances of NLEQ.

We prefer another, more computationally oriented way of selecting the next approximate Jacobian from $Q(y_k, s_k)$. The idea is simple, geometric, and elegant. When we reach step (iv), we have just made a successful step with some approximate Jacobian A_k. It seems reasonable to retain as much of that information contained in A_k as is consistent with the new secant information.

Let $||M||_F = (\sum_{i,j} m_{ij}^2)^{\frac{1}{2}}$ denote the Frobenius norm of a matrix.

Theorem 4.1. Let $s \neq 0$ and y be real n-vectors and let $Q(y,s) = \{Q : Q \text{ is an nxn real matrix and } Qs = y\}$. If A is any nxn matrix then $A_+ = A + \frac{(y - As)s^T}{s^T s}$ is the unique solution to: $\min ||A - Q||_F$ for $Q \in Q(y,s)$.

This theorem is easy to prove [29] and since $s_k \neq 0$, motivates Broyden's [9] method: $A_{k+1} = A_k + \frac{(y_k - A_k s_k)s_k^T}{s_k^T s_k}$

Powell [59], [60] published a general NLEQ routine based on a quasi-Newton algorithm with a Broyden secant approximation to the Jacobians in step (iv). To the best of our knowledge, modified versions of that program have been the best general purpose routines ever since. J. J. Moré of Argonne Labs has a clean simplified version which starts with A_0 gotten from a finite difference or discrete approximation to $J(x_0)$ based on [16].

There is a Kantorovich theorem for Broyden's method in [21] but the following theorem from [12] more accurately validates the excellent performance of Broyden's algorithm.

Theorem 4.2. Let $F : \Omega \subset \mathbb{R}^n \to \mathbb{R}^n$ have a zero x^* in Ω and assume that J is Lipschitz continuous on Ω and that $J(x^*)$ is invertible. There exist positive numbers ε and δ such that if $||x_0 - x^*|| < \varepsilon$ and $||A_0 - J(x^*)|| < \delta$ then Broyden's method starting from x_0 and A_0 is defined and converges to x^*.

Furthermore, $\lim_k e_{k+1}/e_k = 0$ and $\lim_k e_k/||s_k|| = 1$.

The error analysis for this curious method is very interesting. The condition that the quotient of successive error norms go to zero is called "q-superlinear convergence". The error estimate furnished by the norm of the current step is a very trivial consequence of superlinear convergence [28]. A rather surprising fact is that Broyden's method achieves this fast convergence without being consistent i.e., $< A_k >$ does not necessarily converge, much less to $J(x^*)$. The key to the behavior of $< A_k >$ is the following theorem from [28] which characterizes superlinear convergence.

Theorem 4.3. Let F and x^* satisfy the hypothesis of Theorem 4.2 except that J need only be continuous and not Lipschitz continuous. Let $< B_k >$ be any sequence of nonsingular matrices and suppose that for some x_0 the iteration sequence

$$x_{k+1} = x_k - B_k^{-1} F(x_k), \; k = 0,1,\ldots$$

remains in Ω and $x_k \neq x^*$, for $k \geq 0$. Then $< x_k >$ converges q-superlinearly to x^* if and only if

$$\lim_k \frac{[B_k - J(x^*)](x_{k+1} - x_k)}{||x_{k+1} - x_k||} = 0.$$

Recently Gay [36] proved the following completely surprising result about the speed of convergence of $< x_k >$.

<u>Theorem 4.4</u>. Let F and x^* satisfy the hypothesis of Theorem 4.2. Then in addition to the conclusion of Theorem 4.2, there exists a positive constant λ such that for every $k \geq 0$, $e_{k+2n} \leq \lambda e_k^2$.

4.2 <u>Hessian Approximations</u>

We will return to a consideration of NLEQ in the next section but let us now consider UCMIN. In this case it would seem silly not to use a sequence of symmetric approximations to the Hessian matrices in carrying out a quasi-Newton iteration. This would allow us to save storage, it would be useful in step (i) and, if done with care, would probably give a better approximate Hessian since

$$||\tfrac{1}{2}(A_k + A_k^T) - \nabla^2 f(x_k)||_F \leq ||A_k - \nabla^2 f(x_k)||_F.$$

Thus, averaging any approximation A_k with its transpose gives a symmetric Hessian approximation which is guaranteed to be no worse.

Let us suppose then that $A_k \in S \equiv \{M : M \text{ is a real symmetric } n \times n \text{ matrix}\}$ and that this is part of the information we wish to preserve in A_{k+1} from A_k.

Theorem 4.5. Let $A \in S$ and let $x \neq 0$, y be n-vectors. Then $A_+ = A + \frac{(y - As)s^T + s(y - As)^T}{s^Ts} - \frac{s^T(y - As)ss^T}{(s^Ts)^2}$ is the unique solution to $\min ||A - Q||_F$ for $Q \in S \cap Q(y,s)$.

This theorem is easy to prove [29], and since $s_k \neq 0$, it motivates the symmetric Broyden method of Powell [61]

$$A_{k+1} = A_k + \frac{(y_k - A_ks_k)s_k^T + s_k(y_k - A_ks_k)^T}{s_k^Ts_k} - \frac{s_k^T(y_k - A_ks_k)s_ks_k^T}{(s_k^Ts_k)^2}$$

where $A_0 = A_0^T$. Again we refer elsewhere [23] for a Kantorovich Theorem and state the following local result from [12].

Theorem 4.6. Let the hypothesis of Theorem 4.2 hold and assume in addition that $J(x^*)$ is symmetric. If A_0 is chosen to be symmetric, then the conclusions of Theorem 4.2 hold for the symmetric Broyden method.

Near a solution to the UCMIN problem, it is reasonable to assume that F is strictly convex. In addition, there are technical reasons associated with step (ii) why it is useful to maintain positive definite Hessian approximations. Obviously this would be useful in step (i).

We will derive our next two methods by choosing approximations from $Q(y,s) \cap S^+$, where S^+ is the subset of S whose members are positive definite. This time, we will also use the structure of the problem in measuring the change to be made to the current approximation. Any member of S^+ induces an inner product norm on $\mathbb{R}^n$ and the norm $(x^T\nabla^2 f(x^*)x)^{\frac{1}{2}}$ is sometimes referred to as the natural norm induced by UCMIN since level sets of the local quadratic approximation to f are circles in this norm. It would seem reasonable to try to choose A_{k+1} to make the least possible change to A_k measured by $||\nabla^2 f(x^*)^{-\frac{1}{2}}[A_k - A_{k+1}]\nabla^2 f(x^*)^{-\frac{1}{2}}||_F$ but, of course, we don't know the Hessian. Even so, the following theorem from [29] says we can almost do what we wish and in practice, this turns out to be sufficient.

Theorem 4.7. Let $A \in S^+$, s and y be n-vectors. If $s^T y > 0$ then $S^+ \cap Q(y,s) \neq \{\ \}$. Let $M \in S^+$, then for $v = M^{-2}s$,

$$A_+ = A + \frac{(y - As)v^T + v(y - As)^T}{v^T s} - \frac{s^T(y - As)vv^T}{(v^T s)^2}$$

is the unique solution to

$$\min ||M(A - Q)M||_F,\ Q \in Q(y,s) \cap S^+.$$

Now, since the symmetric positive definite matrix $\nabla^2 f(x^*)^{-1}$, has a square root in S^+, we see that all we would really need to minimize the approximate Hessian change in the natural metric would be $v_k = \nabla^2 f(x^*)s_k$, The Davidon [18]-Fletcher-Powell [35] method uses $v_k = y_k$ which can also be looked at as taking M^{-2} to be an arbitrary element of $S^+ \cap Q(y,s)$. This was the first method of the type we are considering and it really opened up the whole field. For definiteness we repeat that the Davidon-Fletcher-Powell or DFP method generates a sequence of symmetric positive definite approximate Hessians $\langle A_k \rangle$ by $A_{k+1} = A_k$ unless $y_k^T s_k > 0$ in which case

$$A_{k+1} = A_k + \frac{(y_k - A_k s_k) y_k^T + y_k (y_k - A_k s_k)^T}{y_k^T s_k} - \frac{s_k^T (y_k - A_k s_k) y_k y_k^T}{(y_k^T s_k)^2} .$$

At the beginning of this section we mentioned that in the real secant method, it was possible to compute the approximation to the reciprocal of the derivative directly. Indeed, it is possible with the same work, to update approximations to the inverse of the Jacobian matrices in the Broyden methods we have discussed. The importance of this is that in either Jacobian or inverse Jacobian form, we are now accomplishing step (iv) with $0(n^2)$ arithmetic operations but if we obtain A_{k+1} then $0(n^3)$ operations are needed for step (i) while only $0(n^2)$ would be needed to obtain s_k^N if we maintain $\langle A_k^{-1} \rangle$.

It was common until a few years ago to maintain $\langle A_k^{-1} \rangle$ for just the reasons above, even though we often had troubles, especially in step (ii), because of numerical singularity in the approximate derivatives. Modern routines maintain $\langle A_k \rangle$ but in factored form, so that step (i) can be carried out in $0(n^2)$ operations, but so that there is greater control over the conditioning of $\langle A_k \rangle$. P. E. Gill and W. Murray of NPL in Britain seem to deserve most of the credit for this practice. See [15], [37], [38], [40], [70]. Moré's program updates a QR factorization of the approximate Jacobians. A recent thesis by Sorensen [70] opens the way for maintaining a symmetric indefinite factorization in the symmetric Broyden method. Following [38] we could maintain an LDL^T decomposition in the DFP method.

Even if we no longer think in terms of generating the sequence $< A_k^{-1} >$ it turns out that the currently preferred secant method for the UCMIN problem can be readily seen from Theorem 4.7 (with y and s interchanged) to solve the following minimization problem:

$$\min \| \nabla^2 f(x^*)^{\frac{1}{2}} [A_k^{-1} - Q^{-1}] \nabla^2 f(x^*)^{\frac{1}{2}} \|_F \text{ for, } Q \in S^+ \cap Q(y_k, s_k)$$

for $y_k^T s_k > 0$. This method is due to Broyden [10], Fletcher [33], Goldfarb [39] and Shanno [69] and is often called the BFGS or complementary DFP method. It is best implemented [38] in the LDL^T form but best understood in terms of Theorem 4.7 as : $A_{k+1}^{-1} = A_k^{-1}$ unless $y_k^T s_k > 0$ in which case

$$A_{k+1}^{-1} = A_k^{-1} + \frac{(s_k - A_k^{-1} y_k) s_k^T + s_k (s_k - A_k^{-1} y_k)^T}{y_k^T s_k} - \frac{y_k^T (s_k - A_k^{-1} y_k) s_k s_k^T}{(y_k^T s_k)^2} .$$

Here is a mystery. Why is the BFGS better in practice than the DFP while the methods due to Broyden [9] and to Greenstadt [43] that solve

$$\min \| A_k^{-1} - Q^{-1} \|_F, \; Q \in Q(y,s)$$

and, for A_k symmetric,

$$\min \| A_k^{-1} - Q^{-1} \|_F, \; Q \in Q(y,s) \cap S$$

are lousy although the Broyden methods are good? Theorems 4.2 and 4.5 hold for these latter two methods by the way [12].

We conclude this section with a theorem from [12].

Theorem 4.8. Let the hypothesis of Theorem 4.2 hold for the UCMIN problem and assume that $J(x^*) = \nabla^2 f(x^*)$ is positive definite as well as symmetric. If A_0 is chosen to be symmetric and positive definite then the conclusions of Theorem 4.2 hold for the DFP and BFGS methods.

5. More Least Change Secant Methods

There are other reasonable choices of M to be made in Theorem 4.7. All the obvious ones yield methods which have been discovered and named. We have only given the ones that seem to be in most common usage. In our opinion, the least change point-of-view makes this group of strange ad-hoc looking methods seem quite reasonable. It certainly is the case that none of them were first discovered in that manner except for those given by Greenstadt [43]. On the other hand, these ideas were at the very heart of the convergence analysis given in [12].

In the next two sections, we would like to take a different tack. The norms used to measure the change from A_k to A_{k+1} will be those we are familiar with, $M = I$ and $M \approx \nabla f(x^*)^{\pm\frac{1}{2}}$, but we will try to incorporate more structure into $< A_k >$ by imposing additional constraints on the set from which A_{k+1} will be chosen.

5.1. Sparsity

It is often the case that $f^i(x)$ in NLEQ will only depend on a few of the components of x. Thus $J(x)$ will have many entries which will always be zero. If n is large, it may be necessary to use sparse matrix techniques in step (i) in order to make the iteration tractable. If we were to use Broyden's method, then $< A_k >$ would be a sequence of full matrices. Schubert [68] and Broyden [11] suggested the following update scheme for which they report good results.

Let $Z = \{A : A$ is a real nxn matrix with $a_{ij} = 0$ if $J(x)_{ij} = 0$ for every $x \in \Omega\}$. Let e_j be the jth column of I. Let a^+ be generalized inverse notation in the sense that $0^+ = 0$ and $a^+ = 1/a$ for $a \neq 0$. The last piece of notation we need is that for any n-vector v, $z_i(v)$ will be the n-vector obtained from v by setting $v^j = 0$ for every j such that $a_{ij} = 0$ for every $A \in Z$. Let $A_k \in Z$ and set $A_{k+1} = A_k + \sum_{j=1}^{n} (||z_j(s_k)||_2^2)^+ e_j e_j^T (y_k - A_k s_k) z_j(s_k)^T$. A_{k+1} is readily shown to be in $Z \cap Q(y_k, s_k)$. Marwil [50] has shown that A_{k+1} is a least change secant method.

Theorem 5.1. Let $s \neq 0$, y be n-vectors and $A \in Z$. Then $A_+ = A + \sum_{j=1}^{n} (||z_j(s)||_2^2)^+ e_j e_j^T (y - As) z_j(s)^T$ is the unique solution to

$$\min ||A - Q||_F \text{ for } Q \in Z \cap Q(y,s).$$

This justifies the name which, with all due respect to Schubert, we suggest: the sparse Broyden method. Marwil [50] proved the following theorem.

Theorem 5.2. Let the hypothesis of Theorem 4.2 hold and let $A_0 \in Z$. Then the conclusions of Theorem 4.2 hold for the sparse Broyden method and $A_k \in Z$ for $k = 1,2,\ldots$.

Neither Schubert nor Broyden used our approach to find the sparse Broyden method. Both Toint [73] and Marwil [50] did find the sparse symmetric Broyden method by solving, for $A_k \in Z$

$$\min ||A_k - Q||_F \text{ for } Q \in Z \cap S \cap Q(y_k, s_k).$$

The formula is somewhat complicated so we won't give it but Toint [74] reports promising computational results.

An important problem is to find a sparse DFP or BFGS method: Let $M^2 \in Z \cap Q(y,s) \cap S^+$ and $A \in Z \cap S^+$ and solve $\min ||M^{-1}[A - Q]M^{-1}||_F$, or $\min ||M[A^{-1} - Q^{-1}]M||_F$ for $Q \in Z \cap S^+ \cap Q(y,x)$.

Can anything be done about maintaining factorizations? This is a very important research question.

5.2 A Partially Available Jacobian

Time and again we have seen NLEQ examples arise in practice in which a significant portion of the Jacobian is available. In other words, there is a natural splitting $J(x) = C(x) + A(x)$ where $C(x)$ can be computed or is ameniable to some special approximation rule and $A(x)$ is unavailable. It seems likely that we have the ideas at hand to build special purpose methods to exploit this situation. After all, Section 4 was the special case $C(x) = 0$.

As a broad outline, let us look at the _default_ technique. We think of step (iv) as consisting of two parts. Let

$B_k = C(x_k) + A_k$ be our approximation to $J(x_k)$. First we obtain $C(x_{k+1})$ and then for some $M \in S^+$ (maybe $M = I$) we obtain A_{k+1} as the solution to :

$$\min ||M[A_k - Q]M||_F \text{ for } Q \in Q(y_k - C(x_{k+1})\, s_k, s_k)$$

which is the same, for $B_{k+\frac{1}{2}} = C(x_{k+1}) + A_k$ as obtaining B_{k+1} as the solution to:

$$\min ||M[B_{k+\frac{1}{2}} - Q]M||_F \text{ for } Q \in Q(y_k, s_k).$$

Of course, one can add further constraints, like symmetry or sparseness.

In order to test these ideas, and to try to solve an important practical problem, we considered the NL2 problem [26]. In that case, $J(x) = \nabla^2\phi(x)$, $C(x) = J(x)^T J(x)$, and $A(x) = \sum_1^n r^i(x)\nabla^2 r^i(x)$. We were working in a context where $A(x)$ could usually not be ignored and so it seemed ideal. Others [2], [3] have tried update methods for approximating $A(x)$ but our least change approach seems to perform better. The final algorithm NL2SOL explained in [26] is a DFP method for NL2. We choose

$$A_{k+1} = A_k + \frac{(y_k^\# - A_k s_k)y_k^T + y_k(y_k^\# - A_k s_k)^T}{y_k^T s_k} - \frac{s_k^T(y_k^\# - A_k s_k)y_k y_k^T}{(y_k^T s_k)^2}$$

where $A_k \in S$ and $y_k = \nabla\phi(x_{k+1}) - \nabla\phi(x_k)$ and $y_k^\# = J(x_{k+1})^T F(x_{k+1}) - J(x_k)^T F(x_{k+1})$. The matrix A_{k+1} solves, for $M^{-2} \in S^+ \cap Q(y_k, s_k)$,

$$\min ||M[A_k - Q]M||_F \text{ for } Q \in Q(y_k^\#, s_k) \cap S.$$

We derived $y_k^\#$ from considering $A(x_{k+1})s_k$ and making the secant approximations $\nabla^2 r^i(x_{k+1})s_k \approx \nabla r^i(x_{k+1}) - \nabla r^i(x_k)$. This performed better than the default which is to choose $A_{k+1} \in Q(y_k - J(x_{k+1})^T J(x_{k+1})s_k , s_k) \cap S$.

5.3 Not For Every Problem But For Some Others

In order for the techniques of 5.2 to be useful it seems that $C(x)$ must represent a nontrivial portion of $J(x)$. It will probably not be possible to carry along a factored sequence of approximate Jacobians using the partial update approach and so step (i) will cost $O(n^3)$ operations as it would for a full Newton method rather than the $O(n^2)$ needed for the methods of Section 4 in which $C(x)$ is neglected. The research questions here are many and obvious.

Frequently in sparse problems, $J(x)$ is either itself available or more likely, efficiently obtainable by finite differences using the clever technique of [16]. In such cases, the expense in step (i) is in obtaining the matrix factorization. Cline [15] has suggested an LU updating technique and Marwil [50] has analyzed a technique in which L is left fixed for several iterations while a sparse Broyden update of U_k into $Q(L^{-1}y_k, s_k)$ is done to obtain U_{k+1}. The k+1 st approximate Jacobian is then thought of as LU_{k+1}. If this idea works out in practice, and preliminary tests indicate that it will, not only would the algorithm have important applicability in its own right, but a whole range of interesting possibilities open up such as updating only the Σ_k part of the SVD $U\Sigma_k V^T$ of A_k, or the bidiagonal part of the first stage of the Golub-Reinsch SVD algorithm, until ill-conditioning forces a restart.

Finally Davidon [19] presents some new interesting ideas investigated further and isolated by Mei [52]. In [67], Schnabel shows their relationship to least change secant methods. Davidon and Nazareth are apparently enjoying success in implementing and testing these ideas at Argonne Labs.

6. Improving A Poor Initial Guess

In Sections 4 and 5 we saw an area in which there is much activity and in which the directions seem clear. Step (iv) of the general quasi-Newton algorithm is in pretty good shape. On the other hand there is some confusion but a good deal of activity in step (ii). How do we improve a poor x_0 to obtain a point at which $x_{k+1} = x_k + s_k^N$ will suffice? There is no important loss of generality for our purposes in restricting ourselves to UCMIN.

Line searches are by far the most commonly used techniques. These are implementations of the theoretical result that tells us to expect convergence for a convex f if we choose $x_{k+1} = x_k + t_k^* s_k^N$ where $f(x_{k+1}) = \min_t f(x_k + ts_k^N)$. This is just a case of trading an n-dimensional problem we can't solve for a sequence of 1-dimensional problems we also can't solve. See [30], [31] or [62] for a relevant discussion of the secant methods.

Up until about ten years ago it was considered not unreasonable to make an honest attempt to find a close approximation to a value of t_k^*. Elaborate and expensive techniques such as bisection and Fibonacci search were employed. None of these routines are still competitive. Basically, modern line search routines settle for any t_k which will make $f(x_{k+1}) < f(x_k)$. They count on finite precision arithmetic to keep t_k from becoming so small that $< x_k >$ converges to a non-minimizer. The essence is in the choice of which value to try for t_k if the previous value wasn't acceptable. The first value tried is virtually always $t_k = 1$ for obvious reasons. As is so often the case in numerical analysis, these methods range from ad hoc to elegant with none clearly the most effective.

It would be pointless to discuss specific examples of searches here, but there is a very interesting open question. Powell [64] gives a fine convergence theorem for the BFGS with an implementable line search. What properties does the sequence $< A_k >$ have to have for this theorem to go through? In particular does it hold for the DFP method? A negative result would be very interesting. Any example of an implementable line search for which a BFGS but not a DFP convergence theorem held would be well received. See [31]. Some other relevant papers are [14], [71].

The line search routines sometimes fail prematurely because the particular direction they are searching in is not a good one. It seems reasonable to change directions as we change t_k. For example, when a very short step is indicated it might be better to take it along $-\nabla f(x_k)$ rather than the Newton direction. This is the effect of the Levenberg [48]-Marquardt, [49], Goldfeldt-Quandt-Trotter [41] idea. Steps (i) and (ii) are combined and $\mu \geq 0$ is the search parameter in seeking an acceptable step $s_k(\mu)$ defined by

$$(\mu I + A_k) s_k(\mu) = -\nabla f(x_k).$$

It turns out that for any $\mu \geq 0$, $s_k(\mu)$ solves the constrained problem

$$\min \left[q_k(x_k + s) \equiv f(x_k) + \nabla f(x_k)^T s + \frac{1}{2} s^T A_k s\right]$$

for $||s||_2 \leq ||s_k(\mu)||_2$.

A very appealing way to use this fact is to think of the quadratic q_k as a model of the objective function f which is accurate in some neighborhood of x_k. One carries along an estimate of a radius ρ_k of this trust region and attempts to stay within the region by picking $\mu = 0$ or so that $||s_k(\mu)||_2 \approx \rho_k$. Since each $s_k(\mu)$ is expensive, the practicality of this method depends on a good scheme for approximately solving the nonlinear equation $g(\mu) = ||s_k(\mu)||_2 - \rho_k = 0$. Moré [54] gives a very clear exposition of an effective implementation based on [44]. Earlier Powell [59], [60], [61] had suggested the dogleg, a segmented approximation to $s_k(\mu)$, which avoided the problem of solving $g(\mu) = 0$. The reader will find an analysis of the dogleg in [63] and an exposition in [27].

The weakness of the model-trust region approach is in the setting of the initial trust radius ρ_0. If $A_0 = I$, as is often the case for the least change secant methods, then there is no rational choice of ρ_0. Experiments [52] show that the algorithm will recover from a poor ρ_0 but at a cost. Currently there is interest in using a line search to help start the trust region. This calls for experiment and analysis. If A_0 is a good approximation to $J(x_0)$ the $\rho_0 = ||s_0(0)||_2 = ||s_0^N||_2$ seems to work pretty well.

As we mentioned previously, the Hessian of f is usually

positive definite near x^*. It has been common practice to maintain $< A_k >$ as a positive definite sequence since this at least had the advantage of making s_k^N a descent direction. Some interesting ideas [51], [55], [70] seem to be emerging for dealing with negative curvature. When x_0 is poor, an indefinite Hessian is not uncommon and so this work has a lot of potential.

There is steady activity in methods based on deforming the problem that must be solved into one for which x_0 is a very good initial guess. The solutions of the deformed problems are then tracked back as the deformation is relaxed. See [4], [5], [6], [7], [17] and also [32] and [46]. Boggs in his thesis traces the idea back to the mid 1800's. The current vogue probably dates to [17].

This area needs more work. The computational results are mixed. There seem to be problems when the solution curve of the parameterized problems cross a manifold on which $J(x)$ drops rank.

Acknowledgement: The author wishes to thank Charles Van Loan for carefully reading this manuscript and suggesting several clarifications. Concerning the unclear points that remain, we absolve him by admitting that he suggested more clarifications that we made.

REFERENCES

1. Abbott, J. P. and Brent, R. P. (1976), "Fast local convergence with single and multistep methods for nonlinear equations." To appear in J. Austral. Math. Soc. (Ser. B).

2. Bartholomew-Biggs, M. C. (1977), "The estimation of the Hessian matrix in nonlinear least squares problems with non-zero residuals," Math. Prog. 12, pp. 67-80.

3. Betts, J. T. (1976), "Solving the nonlinear least square problem: Application of a general method," J.O.T.A. 18, pp. 469-484.

4. Boggs, P. T. (1971), "The solution of nonlinear operator equations by A-stable integration techniques," SIAM J. Numer. Anal. 8, pp. 767-785.

5. Boggs. P. T. (1976), "The convergence of the Ben-Israel iteration for nonlinear least squares problems," Math. Comp. 30, pp. 512-522.

6. Boggs, P. T. (1977), "An algorithm, based on singular perturbation theory, for ill-conditioned minimization problems," SIAM J. Numer. Anal. 14, pp. 830-843.

7. Boggs, P. T. and Dennis, J. E., Jr. (1976), "A stability analysis for perturbed nonlinear iterative methods," Math. Comp. 30, pp. 1-17.

8. Brodlie, K. W. (1977), "Unconstrained minimization," in The State of the Art in Numerical Analysis edited by D. Jacobs, Academic Press.

9. Broyden, C. G. (1965), "A class of methods for solving nonlinear simultaneous equations," Math. Com. 19, pp. 577-593.

10. Broyden, C. G. (1971), "The convergence of a class of double-rank minimization algorithms," Parts I and II, J. Inst. Math. Appl. 6, pp. 76-90, 222-236.

11. Broyden, C. G. (1971), "The converengence of an algorithm for solving sparse nonlinear systems," Math. Comp. 25, pp. 285-294.

12. Broyden, C. G., Dennis, J. E. and Moré, J. J. (1973), "On the local and superlinear convergence of Quasi-Newton methods," J.I.M.A. 12, pp. 223-246.

13. Bryan, C. A. (1968), "Approximate solutions to nonlinear integral equations," SIAM J. Numer. Anal. 5, pp. 151-155.

14. Burmeister, W. (1973), "Die Konvergenzardnung des Fletcher-Powell-Algorithmus," ZAMM 53, pp. 693-699.

15. Cline, A. K. (1976), "A descent method for the uniform solution of over-determined systems of linear equations," SIAM J. Numer. Anal. 13, pp. 293-309.

16. Curtis, A., Powell, M. and Reid, J. (1974), "On the estimation of sparse Jacobian matrices," J. Inst. Math. Appl. 13, pp. 117-120.

17. Davidenko, D. F. (1953), "On a new method of numerical solution of systems of nonlinear equations," Doklady Akad. Nauk. SSSR(N.S.) 88, pp. 601-602.

18. Davidon, W. C. (1959), "Variable metric method for minimization," Argonne Nat. Labs, report ANL-5990 Rev.

19. Davidon, W. C. (1975), "Optimally conditioned algorithms without line searches," Math. Prog. 9, pp. 1-30.

20. Dennis, J. E., Jr. (1968), "On Newton-like methods," Math. 11, pp. 324-330.

21. Dennis, J. E., Jr. (1971), "On the convergence of Broyden's method for nonlinear systems of equations," Math. Comp. 25, pp. 559-567.

22. Dennis, J. E., Jr. (1971), "Toward a unified convergence theory for Newton-like methods" in Nonlinear Functional Analysis and Applications edited by L. B. Rall, Academic Press, New York.

23. Dennis, J. E., Jr. (1972), "On some methods based on Broyden's secant approximation to the Hessian," in Numerical Methods for Non-linear Optimization edited by F. A. Lootsma, Academic Press, London.

24. Dennis, J. E. (1976), "A brief survey of convergence results for quasi-Newton methods," in Nonlinear Programming edited by Cottle, R. and Lemke, C. SIAM-AMS, Proc. Vol. IX.

25. Dennis, J. E., Jr. (1977), "Nonlinear least squares and equations," in The State of the Art in Numerical Analysis edited by D. Jacobs, Academic Press.

26. Dennis, J. E., Gay, D. M. and Welsch, R. E. (1977), "An adaptive nonlinear least-squares algorithm," Cornell Computer Science TR77-321, (submitted for publication).

27. Dennis, J. E. and Mei, H. W. (1975), "An unconstrained optimization algorithm which uses function and gradient values," Cornell Computer Science TR77-246, (to appear in J.O.T.A.).

28. Dennis, J. E. and Moré, J. J. (1974), "A characterization of superlinear convergence and its application to quasi-Newton methods," Math. Comp. 28, pp. 549-560.

29. Dennis, J. E. and Moré, J. J. (1977), "Quasi-Newton methods, motivation and theory," SIAM Review 19, pp. 46-89.

30. Dixon, L. C. W. (1972a), "Quasi-Newton family generate identical points," Math. Prog. 2 pp. 383-387.

31. Dixon, L. C. W. (1972b), "The choice of step length, a crucial factor in the performance of variable metric algorithms," in Numerical Methods for Non-linear Optimization edited by F. A. Lootsma, Academic Press, London.

32. Eaves, B. C. (1976), "A short course in solving equations with PL homotopies," in Nonlinear Programming edited Cottle, R. and Lemke, C. SIAM-AMS Proc. Vol. IX.

33. Fletcher, R. (1970), "A new approach to variable metric algorithms," Comput. J. 13, pp. 317-322.

34. Fletcher, R. (1977), "Methods for solving non-linearly constrained optimization problems," in The State of the Art in Numerical Analysis edited by D. Jacobs, Academic Press.

35. Fletcher, R. and Powell, M. J. D. (1963), "A rapidly convergent descent method for minimization," Comput. J. 6, pp. 163-168.

36. Gay, D. M. (1979), "Some convergence properties of Broyden's method," SIAM J. Numer. Anal. 16, pp. 623-630.

37. Gill, P. E. Golub, G., Murray, W., Saunders, M. A. (1974), "Methods for modifying matrix factorizations," Math. Comp. 28, pp. 505-536.

38. Gill, P. E. and Murray, W. (1972), "Quasi-Newton methods for unconstrained minimization," J.I.M.A. 9, pp. 91-108.

39. Goldfarb, D. (1970), "A family of variable-metric methods derived by variational means," Math. Comp. 24, pp. 23-26.

40. Goldfarb, D. (1976), "Factorized variable metric methods for unconstrained optimization," Math. Comp. 30, pp. 796-811.

41. Goldfeldt, S. M., Quandt, R. E. and Trotter, H. F. (1966), "Maximization by quadratic hill-climbing," Econometrica 34 pp. 541-551.

42. Golub, G. H. and Pereyra, V. (1973), "The differentiation of pseudoinverses and nonlinear least squares problems whose variables separate," SIAM J. Numer. Anal 10, pp. 413-432.

43. Greenstadt, J. (1970), "Variations on variable-metric methods," Math. Comp. 24, pp. 1-18.

44. Hebden, M. D. (1973), "An algorithm for minimization using exact second derivatives," A.E.R.E. Harwell Rpt. T.P. 515.

45. Kaufman, L. (1975), "A variable projection method for solving separable nonlinear least squares problems," BIT 15, pp. 49-57.

46. Kellogg, R., Lui, T. and Yorke, J. (1976), "A constructive proof of the Brouwer Fixed Point Theorem and computational results," SIAM J. Numer. Anal. 13, pp. 473-483.

47. Krogh, F. T. (1974), "Efficient implementation of a variable projection algorithm for nonlinear least squares problems," Comm. ACM 17, pp. 167-169.

48. Levenberg, K. (1944), "A method for the solution of certain nonlinear problems in least squares," Quart. Appl. Math 2, pp. 164-168.

49. Marquardt, D. W. (1963), "An algorithm for least squares estimation of nonlinear parameters," SIAM J. Appl. Math. 11, pp. 431-441.

50. Marwil, E. S. (1978), "Exploiting sparsity in Newton-like methods," Ph.D. thesis, Cornell University.

51. McCormick, G. P. (1977), "A modification of Armijo's step-size rule for negative curvature," Math. Prog. 13, pp. 111-115.

52. Mei, H. H. W. (1977), "An analysis and implementation of Davidon's techniques for unconstrained optimization," Ph.D. thesis, Cornell University.

53. Meyer, Gunter H. (1968), "On solving nonlinear equations with a one-parameter operator imbedding," SIAM J. Numer. Anal. 5, pp. 739-752.

54. Moré, J. J. (1978), "The Levenberg-Marquardt algorithm: implementation and theory," Numerical Analysis, Lecture Notes in Math., No. 630, Springer, Berlin, 1978.

55. Moré, J. J. and Sorensen, D. (1979), "A modified Newton method using directions of negative curvature," Math. Prog. 16, pp. 1-20.

56. Oren, S. S. (1973), "Self-scaling variable metric algorithm without line search for unconstrained minimization," Math Comp. 27, pp. 873-885.

57. Ortega, J. M. and Rheinboldt, W. C. (1970), Iterative Solution of Nonlinear Equations in Several Variables, Academic Press, New York.

58. Osborne, M. R. (1975), "Some special nonlinear least squares problems," SIAM J. Numer. Anal. 12, pg. 571.

59. Powell, M. J. D. (1970a), "A hybrid method for nonlinear equations," in Numerical Methods for Nonlinear Algebraic Equations, P. Rabinowitz, ed., Gordon and Breach, London.

60. Powell, M. J. D. (1970b), "A FORTRAN subroutine for solving systems of nonlinear algebraic equations," in Numerical Methods for Nonlinear Algebraic Equations, P. Rabinowitz, ed., Gardon and Breach, London.

61. Powell, M. J. D., (1970), "A new algorithm for unconstrained optimization," in Nonlinear Programming edited by J.B. Rosen, O.L. Maugasarian, K. Ritter, Academic Press, New York.

62. Powell, M. J. D. (1972), "Some properties of the variable metric method," in Numerical Methods for Nonlinear Optimization, edited by F.A. Lootsma, Academic Press, London.

63. Powell, M. J. D. (1974), "Convergence properties of a class of minimization algorithms," A.E.R.E. Harwell tech. rept. C.S.S.8.

64. Powell, J. J. D. (1975), "Some global convergence properties of a variable metric algorithm for minimization without exact line searches," in Nonlinear Programming, edited by Cottle, R. and Lemke, C., SIAM-AMS, Proc. Vol IX.

65. Powell, M. J. D. (1976), "Algorithms for nonlinear constraints that use Lagrangian functions," presented at Ninth International Symposium on Math. Prog., Budapest.

66. Ruhe, A. and Wedin, P. A. (1974), "Algorithms for separable nonlinear least squares problems," Univ. of Umea, Dept. of Info. Processing, rept. UMINF-47.74.

67. Schnabel, R. B. (1977), "Analyzing and improving quasi-Newton methods for unconstrained optimization", PhD thesis, Cornell University.

68. Schubert, L. K. (1970), "Modification of a quasi-Newton method for nonlinear equations with a sparse Jacabian," Math. Comp. 24, pp. 27-30.

69. Shanno, D. F. (1970), "Conditioning of Quasi-Newton methods for function minimization," Math. Comp. 24, pp. 647-656.

70. Sorensen, D. C. (1977), "Updating the symmetric indefinite factorization with application in a modified Newton method," Argonne National Labs. Rpt. 77-49.

71. Stoer, J. (1975), "On the convergence rate of imperfect minimization algorithms in Broyden's β-class," Math. Prog. 9 pp. 313-335.

72. Tapia, R. A. (1977), "Equivalences of quasi-Newton methods for constrained optimization and a new class of algorithms, Rice University, Dept. of Math Sciences.

73. Toint, Ph. L. (1977), "On sparse and symmetric matrix updating subject to a linear equations," to appear Math. Comp.

74. Toint, Ph. L. (1977), "Some numerical results using a sparse matrix updating formula in unconstrained optimization," Cambridge University, Dept. of App. Math. and Th. Phys. Report DAMTP 77/NA4.

75. Wedin, P. A. (1974b), "On surface dependent properties of methods for separable nonlinear least squares problems," Inst. for tellampad matematik, Box 5073, Stockholm 5, ITM Arbetsrapport nr. 23.

Appendix A. Matrix Factorizations

The schedule for the AMS Shortcourse for which the preceeding paper was prepared, calls for the material on nonlinear optimization to be presented after Cleve Moler lectures on numerical linear algebra. Thus, the main body of the paper doesn't go into details on the computation of the Newton step by solving the linear system in step (i) of the quasi-Newton algorithm of section 2.

This omission is made partly for brevity and partly to avoid any confusion that might arise from slight differences in notation. The purpose of this appendix is to provide enough of an outline to make the preceeding paper somewhat self-contained. More details can be found in Introduction to Matrix Computation by G. W. Stewart III, Academic Press, 1973.

Mathematics students taking their first numerical analysis course are often surprised to learn that computing A^{-1} and then $A^{-1}b$ is a bad way to solve $Ax = b$. Numerical methods usually involve decomposing A into the product of from two to four matrices. The form of the decomposition depends on properties of A and the computing environment but the candidate factors should be so that it is easy to solve a linear system with one of the factors as coefficient matrix. In general, the factorization, say $A = A_1 \cdot A_2 \cdot A_3$ is the first step. One then thinks of successively solving $A_1 x_1 = b$ for x_1 and then $A_2 x_2 = x_1$ for x_2 and finally $A_3 x = x_2$ for x. As a practical matter, some of the intermediate linear systems are often solved as the factorization stage proceeds.

As an illustration, let us consider the traditional form of

Gaussian elimination in which rows (or equations) are interchanged if a zero appears on the main diagonal of the upper triangular matrix being generated. This algorithm corresponds to a factorization $A = P^T LU$ where P is a permutation matrix and L and U are lower and upper triangular respectively. But the solution to $P^T x_1 = b$, $x_1 = Pb$ as well as the solution x_2 to $Lx_2 = x_1 = Pb$ are often generated as the factorization stage proceeds. This leaves the system $Ux = x_2$ as the only obvious "factor" system to be solved. This example really illustrates the use of all the factorizations to solve $Ax = b$ and we complete this appendix by cataloging the factorizations we refer to.

The LU factorization: In fact we use this term for the decomposition

$$PA = LU \text{ or } A = P^T LU$$

where P is a permutation matrix, L is unit lower triangular ($l_{ii} = 1$) and U is upper triangular. This is "Gaussian elimination" and it is intended for general well-conditioned square problems.

The Cholesky factorization: If A is symmetric and positive definite, then $A = LL^T$ can be obtained for about half the work and storage of $A = LU$. Generally in practice one uses the related decomposition $A = LDL^T$ where L is unit lower triangular and D is a positive diagonal matrix.

The symmetric indefinite factorization: In this case, A is only assumed symmetric and the decomposition is $A = PLDL^T P^T$ where P

is a permutation matrix, L is unit lower triangular and D is a block diagonal matrix of 1x1 and 2x2 blocks.

The QR decomposition: In this case, A has no special properties. In fact, it could even be rectangular. We write $A = QRP^T$ where P is a permutation matrix, R is upper triangular and Q is orthogonal, i.e., $Q^TQ = I$.

The SVD or singular value decomposition: Here, A may not only be rectangular but it can even be rank deficient. The decomposition is $A = UDV^T$ where U and V are orthogonal matrices and D is a nonnegative diagonal matrix. The rank of A is the rank of D. The SVD is related to the polar decomposition of A:

$$A = (UV^T)(VDV^T).$$

Proceedings of Symposia in Applied Mathematics
Volume 22
1978

The approximation of functions and linear functionals: Best vs. good approximation

Carl de Boor
University of Wisconsin
Madison, WI 53706

Outline

1. Approximation of functions

   Characterization of a best approximation
   - Least-squares approximation
   - Least-mean approximation
   - Uniform approximation

   Why bother to construct best approximations?
   Example: Polynomial approximation on an interval
   Linear projectors

2. Approximation of linear functionals

   The approaches of Newton, Sard and Golomb-Weinberger
   Hypercircle
   Example: Optimal quadrature rules for $\mathbb{L}_2^{(k)}[a,b]$
   Why are the optimal rules not used?
   Automatic numerical quadrature

1. Approximation of functions

The theory of best approximation from a finite dimensional linear subspace S of a real normed linear space X of functions is very pretty. I discuss it here for its own sake and in order to illustrate the typical concerns of approximation theory. For more detail, see [2], [4], [5].

Let X be a normed linear space with norm $\|\cdot\|$, let S be a subset of X, and let $x \in X$. A best approximation (or, b.a.) to x from S is any element $s^* \in S$ for which

$$\|x - s^*\| = \operatorname{dist}(x, S) := \inf_{s\in S} \|x - s\| .$$

Typical questions asked are: (i) Existence: Does x have a b.a. from S? (ii) Uniqueness: How many b.a.'s from S does x have? (iii) Characterization: How does one recognize a b.a.? (iv) Construction: How does one compute a b.a.? (v) A priori bounds: Given some classifying information about x, what can be said about $\operatorname{dist}(x, S)$?

Let S be, specifically, a finite dimensional linear subspace. Then, existence is guaranteed since S is closed, and bounded sets in S are totally bounded (or, precompact). Uniqueness is certain as long as the norm is strictly convex. Since some norms of practical interest (such as the uniform norm on the continuous functions $C(T)$ on some compact metric space T, or the $\mathbb{L}_1$-norm) fail to be strictly convex, there is a longstanding investigation into the circumstances under which a b.a. from S is unique in spaces in which the unit sphere contains line segments. Best known results here concern Haar spaces in $C(T)$, i.e., subspaces of $C(T)$ of dimension n over which any n linear functionals of point evaluation (at distinct points) are linearly independent.

Practically all characterization theorems are specializations of the following:

Characterization Theorem: $s^* \in S$ is a b.a. from S to x iff there exists a (nontrivial) linear functional λ which vanishes on S and takes its norm on the error $e := x - s^*$. In symbols,

$$\|x - s^*\| = \operatorname{dist}(x, S) \quad \underline{\text{iff}} \quad \lambda(x-s^*) = \|\lambda\|\|x-s^*\| \quad \underline{\text{for}}$$
$$\underline{\text{some}} \quad \lambda \in S^{\perp} \setminus \{0\}.$$

The proof of this theorem consists of an application of the Hahn-Banach Theorem. Note that, if also s' is a b.a. to x from S, then necessarily $\lambda(x-s') = \|\lambda\|\|x-s'\|$ for the same λ, so that questions of nonuniqueness are closely tied to the possibility of having more than one extremal for certain linear functionals on X.

Although this may not always be apparent, algorithms for the numerical construction of a b.a. are usually based on this characterization theorem. We discuss three important cases:

Least-squares approximation. Here, the norm derives from an inner product,

$$\|x\|^2 = \langle x,x \rangle,$$

and the linear functional which takes on its norm on the error is necessarily a scalar multiple of the error itself, taken as a linear functional via the inner product, i.e.,

$$\lambda = \langle \cdot, x-s^* \rangle.$$

This leads to the familiar characterization that s^* is a b.a. from S to x iff the error $x-s^*$ is orthogonal to S, and, with $(s_i)_1^n$ a basis for S, one obtains the so called normal equations

$$\sum_{j=1}^{n} \langle s_i, s_j \rangle a_j = \langle s_i, x \rangle, \quad i=1,\ldots,n$$

which characterize the coefficients of a b.a. $s^* = \Sigma_j a_j s_j$.

Least-mean approximation. Here, $X = \mathbb{L}_1(T;\mu)$, and

$$\|x\| = \int_T |x(t)| \mu(dt).$$

Approximation in such norms has gained recent favor among statisticians because such norms are less sensitive to isolated bad behavior of x ("robust regression" is the catch phrase) than are $\mathbb{L}_2$ or $\mathbb{L}_\infty$ norms. Continuous linear functionals on $\mathbb{L}_1(T;\mu)$ are given as integration against essentially bounded functions on T, i.e., $\lambda x = \int_T h_\lambda(t)x(t)\mu(dt)$, and $\|\lambda\| = \|h_\lambda\|_\infty$, for some appropriate $h_\lambda \in \mathbb{L}_\infty(T;\mu)$. The equality

$$\lambda e = \int_T h_\lambda(t)e(t)\mu(dt) = \|h_\lambda\|_\infty \int_T |e(t)|\mu(dt)$$

to be satisfied by the error $e = x - s^*$ forces

$$h_\lambda(t) = \text{const signum } e(t) \quad \text{for } \mu\text{-a.e. } t \text{ with } e(t) \neq 0,$$

and computational schemes are based on constructing an approximation s^* for which $\text{signum}(x - s^*)$ is orthogonal to S. This is a nontrivial and nonlinear process except in the following happy circumstance: T is an interval, and a signum function with exactly n sign changes, at the points $t_1 < \ldots t_n$ say, is known which is orthogonal to S, <u>and</u>, on interpolating x at these points by an element from S, we discover that the interpolation error changes sign only at the t_i's; then the interpolant is also a best $\mathbb{L}_1$-approximation.

<u>Uniform approximation</u>. Now, $X = C(T)$, the space of continuous functions on some compact metric space T, the norm is

$$\|x\| = \sup_{t\in T} |x(t)| .$$

In this setting, the characterization theorem can be considerably sharpened with the aid of the following lemma. In its statement, I use the notation $[t]$ for the linear functional on $C(T)$ of evaluation at t,

$$[t]x := x(t) .$$

<u>Lemma</u>. <u>Any linear functional μ on some r-dimensional subspace M of $C(T)$ has a norm-preserving extension to all of $C(T)$, of the form $\Sigma_1^r w_i[t_i]$</u>.

In other words, for a given linear functional μ on M, one can find r points $t_1, \ldots, t_r$ and weights (w_i) so that $\mu m = \Sigma_i w_i m(t_i)$ for all $m \in M$ (so much is trivial), but also so that $\Sigma_i |w_i| = \sup_{m\in M} \mu m/\|m\|$. One proves this lemma by showing that the set

$$E := \{a[t]|_M : |a| = 1, t \in T\}$$

is compact, hence so is its convex hull co(E). Further, co(E) is contained in the unit ball B of the dual M^* of M. On the other hand, if $\mu \in M^* \setminus co(E)$, then we can strictly separate it from co(E) by a hyperplane, i.e., there exists then $m \in M = M^{**}$ so that

$$\mu m > \sup_{\lambda \in co(E)} \lambda m \geq \sup_{t \in T} |m(t)| = \|m\|$$

showing that any such μ cannot be in B, either. This proves that $co(E) = B$, hence, by Caratheodory's Theorem, any point in the boundary of B can be written as a convex combination of at most r points in E.

By applying this lemma to the linear functional $\lambda|_M$ of the Characterization Theorem, with $M := \text{span}\,\{x\} \cup S$, one obtains the

Characterization Theorem for $C(T)$: $s^* \in S$ is a b.a. to x from the n-dimensional linear subspace S of $C(T)$ iff there exist points $t_1, \ldots, t_{n+1}$ and coefficients $w_1, \ldots, w_{n+1}$ so that the linear functional $\lambda := \Sigma_i w_i[t_i]$ satisfies

$$\|\lambda\| = 1 \ ; \ \lambda \in S^{\perp} \ ; \ \lambda(x - s^*) = \|x - s^*\| .$$

In effect, it must be possible to construct a linear functional of the form $\Sigma_i w_i[t_i]$ involving at most $n+1$ distinct points, each of which is a (global) extreme point of the error $x - s^*$, with the coefficient w_i nonzero and of the same sign as $(x - s^*)(t_i)$, and so that $\Sigma_i w_i s(t_i) = 0$ for all $s \in S$.

When S is a Haar space, such a linear functional necessarily involves exactly $n+1$ points and, if T is an interval and these points are ordered, $t_1 < \ldots < t_{n+1}$, then $w_i w_{i+1} < 0$, all i. In this way, we then get the wellknown Alternation Theorem: The error $x - s^*$ in a b.a. must take on its extreme value at least $n+1$ times, and with alternating sign. This leads to extrema-equalizing algorithms such as the Remez algorithms for the construction of a b.a. from Haar spaces which, in the case of well understood spaces such as polynomials, are only moderately expensive.

To be a bit more explicit, the (second) Remez algorithm improves an existing approximation $s \in S$ by selecting from the sequence of extreme points of the error $x - s$ a subsequence $t_1 < \dots < t_{n+1}$ so that $\max_i |(x-s)(t_i)| = \|x-s\|$ and $(x-s)(t_i)(x-s)(t_{i+1}) < 0$, all i. Then, it constructs the best approximation to x from S with respect to the discrete norm $\|x\|_t := \max_i |x(t_i)|$. Iteration of this step can be shown to converge, usually quadratically since it can be interpreted as an instance of Newton's method. Even the search for the extreme points of $x-s$ (which could take a lot of work if done too carefully) can be built into this Newton-like iteration.

Unfortunately, even practically important spaces of univariate functions such as spline functions (i.e., piecewise polynomial functions) are not (quite) Haar, and there are no Haar spaces (of dimension > 1) of multivariate functions. Much effort is still being expended to overcome this handicap. Clever and costly algorithms have been proposed and used, based on the above characterization theorem, for the construction of a b.a. in that case.

Why bother to construct best approximations? So far, then, I have shown you (linear) approximation theory as a pretty and cohesive application of basic and simple functional analysis. It gets more interesting when the approximating set S is nonlinear, such as rational functions, or exponential sums, or piecewise polynomial functions with variable breakpoints. But, instead of discussing such matters, I now want to take the role of a user of approximation theory and, in that role, raise the question whether the construction of a best approximation is worth the effort. Clearly, it depends on why a b.a. is being constructed in the first place and on what alternatives are available. In typical problems, one has available not just one approximating set, but a whole scale (S_h) of them, indexed by some discrete or continuous parameter h, which we think of as going to zero. As it does, $\mathrm{dist}(x, S_h)$ also goes to zero, while the dimension of S_h increases. This means that the approximation s^* becomes more complex (more expensive to evaluate) as h decreases. Still, one might prefer a slightly more complex approximation obtained cheaply to a less complex approximation of the same proximity but obtained only at great expense.

Example: Polynomial approximation on an interval. A classical illustration is provided by polynomial approximation on some interval, say on $[-1,1]$, to a function $x \in C[-1,1]$. Let $P_k x$ denote the interpolant to x from $\mathbb{P}_k$, the set of polynomials of order k (or, degree $< k$), at the so-called Chebyshev points

$$t_i := \cos((2i-1)\pi/(2k)), \quad i=1,\ldots,k ,$$

the extrema on $[-1,1]$ of the Chebyshev polynomial of degree k. The interpolation error then has (see next page) the bound

$$\|x - P_k x\|_\infty \leq (1 + \|P_k\|)\operatorname{dist}_\infty(x, \mathbb{P}_k)$$

with the norm $\|P_k\|$ of P_k, as a map on $C[-1,1]$, bounded by

$$\|P_k\| \leq (2/\pi) \ln k + 1 .$$

This bound is sharp in the sense that

$$\lim_{k\to\infty} \|P_k\|/ \ln k = 2/\pi .$$

Consequently, we may have $\|x - P_k x\|_\infty/\operatorname{dist}_\infty(x, \mathbb{P}_k)$ go to infinity with k. In other words, the best polynomial approximation to x from $\mathbb{P}_k$ may well be very much better than this polynomial interpolant. But, from a practical, numerical point of view, this observation is i r r e l e v a n t . It seems extravagant even to contemplate the approximation of some function x by polynomials of order 1000 and, for $k \leq 1000$, we have only $\|P_k\| < 5.4$. In the more realistic range $k \leq 20$, we have $\|P_k\| < 2.9$. This says that, in practically contemplated applications, use of the polynomial interpolant instead of the best possible approximant may cost, at worst, half a decimal digit of accuracy. If the function x is at all well approximable by polynomials, then we can more than make up for this loss by going to a polynomial of slightly higher order. At the same time, the computational cost for constructing a best uniform polynomial approximation is easily ten times that of constructing an interpolant. Except in the special case when the approximation is to be evaluated many, many times, such as in a library routine for the function x, it usually does not pay to construct the best polynomial approximant.

Linear projectors. While one might, offhand, consider use of any map A from X into S as an alternative to the construction of a b.a. from S, special attention has been paid to (linear) interpolation maps, i.e., to linear projectors. Such a map P is characterized by the fact that

$$Px \in S \quad \text{and} \quad \lambda Px = \lambda x \quad \text{for all } \lambda \in \Lambda$$

for some n-dimensional linear space of linear functionals on X with $S \cap \Lambda_{\perp} = \{0\}$. Here, $\Lambda_{\perp} := \{x \in X : \lambda x = 0, \text{ all } \lambda \in \Lambda\}$ is then the kernel of P and S is its range. Because such maps are idempotent, we have, for any $s \in S$, $Ps = s$ and so, since such P is linear, $\|x - Px\| = \|(1 - P)(x - s)\| \leq \|1 - P\| \|x - s\|$. Taking now the infimum over all $s \in S$, we get "Lebesgue's inequality"

$$\|x - Px\| \leq \|1 - P\| \operatorname{dist}(x, S),$$

which relates the interpolation error $\|x - Px\|$ to the best possible error $\operatorname{dist}(x, S)$.

The search for optimal interpolation schemes, i.e., for projectors P onto S for which $\|1 - P\|$ is as small as possible, (as carried out most recently so vigorously by Cheney and his coworkers) has turned out to be very difficult except in the trivial case when $\inf \|1 - P\| = 1$. This is, e.g., the case when X is an inner product space, in which case the optimal interpolant Px to x is simply the b.a. to x from S.

In spaces other than inner product spaces, few optimal projectors have been identified. This is not too surprising since one is dealing here with best approximation, to the identity, from an infinite dimensional manifold. In any case, while it might be nice to know more about optimal projectors, the numerical analyst puts an additional constraint on the problem: It should be relatively inexpensive to construct Px . This severely restricts the choice of the interpolation conditions, i.e., the choice of Λ . In effect, in concrete function spaces X , Λ must be spanned by point evaluations or, perhaps, certain simple linear combinations of point evaluations. This makes, e.g., Least-squares approximation less than acceptable, except in some discretized form.

Polynomial interpolation at the Chebyshev points is an example of a "good" (even though not optimal) interpolant whose construction is inexpensive. There are similar successful examples for spline spaces of one variable. But, at present, woefully little is known about such "good" projectors in multivariate function spaces. How should one choose the $k(k+1)/2$ interpolation points in the unit square $U \subseteq \mathbb{R}^2$ so that the projector of interpolation from $\mathbb{P}_k$:= polynomials in two variables of (total) degree $< k$ is defined and has "small" norm? What about shapes other than the square? If the theory of Gauss quadrature over multivariate domains is any guide, these are tricky questions. On the other hand, construction of best approximations to multivariate functions is very expensive, and such "good" alternatives would be very welcome.

Finally, the preceding discussion gives me an excuse to mention briefly the main service performed by Approximation Theory for Numerical Analysis, viz. the furnishing of a priori bounds. As I said earlier, one usually has a scale (S_h) of approximating sets, indexed by some discrete or continuous h which goes to zero. Correspondingly, one has a scale (P_h) of linear projectors. If this scale is stable, i.e. bounded uniformly in h , then Lebesgue's inequality reduces consideration of the rapidity with which $P_h x$ approaches x as h goes to zero to considerations of the rate at which

$$\mathrm{dist}(x, S_h)$$

goes to zero with h . For standard scales, such as spaces of polynomials, or of piecewise polynomials (splines), Approximation Theory provides a detailed analysis of $\operatorname{dist}(x, S_h)$ as a function of h and as a function of special properties, typically "smoothness" of some kind, which x may have. The code word for such results is "direct", or "Jackson" (after D. Jackson) in distinction to "inverse", or "Bernstein" (after S.N. Bernshtein) which establish the sharpness of direct estimates and so prove that, for certain classes of x, $\operatorname{dist}(x, S_h)$ cannot decrease faster than a certain rate.

Most recently, the finite element theory has been the happy beneficiary and, I must add, important contributor to, this central area of Approximation Theory.

2. Approximation of linear functionals

I am supposed to cover not only Approximation Theory, but also quadrature or, more generally, the approximation of linear functionals. Here, too, I want to give you first a taste of the available mathematical theory, quite pretty, but then discuss what is actually used in computations.

The task of evaluating an integral numerically, or of estimating a derivative numerically, can be given a rigorous abstract form in various ways. In all of them, one assumes as given some information about a certain element g of the linear space X and wishes to deduce from this information some information about μg, with μ a given linear functional on X.

Classically, one assumes as known the values $(\lambda_i g)$ of g at certain linear functionals $\lambda_1, \ldots, \lambda_n$ on X. Following the procedure in vogue since Newton (and still the only one in our textbooks), one would approximate μg by $\mu P g$, with P a convenient linear projector having

$$\Lambda := \operatorname{span}(\lambda_i)_1^n$$

as its space of interpolation conditions. Equivalently, one approximates μg by a "rule" of the form $\Sigma_i w_i \lambda_i g$ which is exact for a certain linear subspace S, viz. the range of P. This leads to the error bound

$$|\mu g - \mu P g| \leq \operatorname{dist}(\mu, \Lambda)\, \|1 - P\|\, \operatorname{dist}(g, S)$$

in case X is a normed linear space and all linear functionals involved are continuous, which shows the possible "double accuracy" of such an approximation process.

<u>Sard</u>'s approach [6] consists in introducing a norm on X with respect to which the λ_i's and μ are continuous, after which he constructs a best approximation to μ from Λ with respect to the corresponding norm on linear functionals. This produces then a rule

$$\mu g \sim \sum_{i=1}^{n} w_i \lambda_i g$$

for the approximation of μg whose worst error (as g ranges over all of X) is as small as possible.

<u>Golomb and Weinberger</u> [3] work in the same setup as Sard but assume that, in addition to the values $(\lambda_i g)$, one also knows some ρ for which

$$\|g\| \leq \rho .$$

(Actually, both Sard and Golomb & Weinberger start with some seminorm v on X for which $(\ker v) \cap \Lambda_{\perp} = 0$. This gives the existence of some linear projector Q with $\ker v$ in its range and $\Lambda_{\perp}$ in its kernel, and X is then normed by

$$\|x\| := \|Qx\|_* + v(x) ,$$

with $\|\cdot\|_*$ any convenient norm on the finite dimensional subspace $\operatorname{ran} Q$. Best approximation to μ in the corresponding norm on linear functionals would produce a rule which is exact on $\operatorname{ran} Q$.)

The problem attacked by Golomb & Weinberger then is the determination of the interval of all possible values for μg given only the numbers $(\lambda_i g)$ and the bound ρ on $\|g\|$. This amounts to calculating the interval μF_ρ, with

$$F_\rho := \{x \in X : \lambda_i x = \lambda_i g,\ i=1,\ldots,n \ ;\ \|x\| \leq \rho\} .$$

Now, this problem (as I learned from H.G. Burchard) can be reinterpreted as an extension problem:

We are given the linear functional $g|_\Lambda$ on the linear subspace Λ of X^* and wish to extend it linearly to the point $\mu \in X^* \setminus \Lambda$ in such a way that the resulting linear functional on

$$M := \operatorname{span} \{\mu\} \cup \Lambda$$

has norm no bigger than ρ. The interval μF_ρ consists exactly of those numbers which we could assign to the extension as its value at μ without raising the norm of the resulting extension to M above ρ. The classical proof of the Hahn-Banach Theorem therefore supplies the computable answer as follows. If the value of the extension at μ is a, then the norm of the extension is

$$\sup_{\alpha \in \mathbb{R}, \lambda \in \Lambda} \frac{|\alpha a + \lambda g|}{\|\alpha\mu + \lambda\|} = \sup_{\lambda \in \Lambda} \frac{|a - \lambda g|}{\|\mu - \lambda\|} .$$

The extension has therefore norm $\leq \rho$ iff

$$\sup_{\lambda \in \Lambda} (\lambda g - \rho\|\mu - \lambda\|) \leq a \leq \inf_{\lambda \in \Lambda} (\lambda g + \rho\|\mu - \lambda\|) .$$

The sup and inf here are to be taken of a concave, resp. convex function over some finite dimensional set which could be taken to be bounded. There is then, in principle, no difficulty in finding them. In practice, though, there is no hope of dealing with this problem unless one is able to represent X^* as some space of functions acting on X via some pairing.

The hypercircle. Golomb and Weinberger considered this problem only in a Hilbert space. In general, the set

$$F := \{x \in X : x|_\Lambda = g|_\Lambda\}$$

is of the form $F = g + \Lambda_\perp$. But, when X is an inner product space, then we can choose a $g^* \in F$ which is orthogonal to $\Lambda_\perp$, i.e., we can write

$$F = g^* + \Lambda_\perp \quad \text{with} \quad g^* \perp \Lambda_\perp$$

Indeed, choose g^* as the unique element of minimal norm in F. Then $\|g^* + h\|^2 = \|g^*\|^2 + \|h\|^2$ for all $h \in \Lambda_\perp$, and therefore

$$F_\rho = \{g^* + h : h \in \Lambda_\perp , \|h\|^2 \le \rho^2 - \|g^*\|^2\},$$

a hypercircle in the nomenclature of Synge . Consequently,

$$\mu F_\rho = \mu g^* + \mu\{h \in \Lambda_\perp : \|h\|^2 \le \rho^2 - \|g^*\|^2\},$$

showing μF_ρ to have midpoint or center μg^* and radius or half length equal to

$$\|\mu|_{\Lambda_\perp}\|(\rho^2 - \|g^*\|^2)^{1/2} .$$

Note that g^* does not depend on μ , it serves as the center of the hypercircle F_ρ .

Even with this simplified description of μF_ρ, we are still not ready for numerical work. For this, we now make use of the fact that we may identify Λ with the subspace $\hat{\Lambda}$ of its representers in X (λ being identified with $\hat{\lambda} \in X$ for which $\lambda = \langle \cdot , \hat{\lambda} \rangle$). Since F is a translate of

$$\Lambda_\perp = \hat{\Lambda}^\perp := \text{orthogonal complement of } \Lambda,$$

F has exactly one point in common with $\hat{\Lambda}$, viz. the point g^*. Consequently, $g^* = Px$ for all $x \in F$, with P the orthogonal projector onto $\hat{\Lambda}$. Note that Λ provides the interpolation conditions for this linear projector P .

Further,

$$\|\mu|\Lambda_\perp\|^2 = \sup_{h\in\Lambda_\perp} \left(\frac{\mu h}{\|h\|}\right)^2 = \sup_{h\in\Lambda_\perp} \frac{\langle h, \hat{\mu}\rangle^2}{\langle h,h\rangle} = \sup_{h\in\Lambda_\perp} \frac{\langle h,(1-P)\hat{\mu}\rangle^2}{\langle h,h\rangle}$$

$$= \|(1 - P)\hat{\mu}\|^2 = \operatorname{dist}(\mu,\Lambda)^2 .$$

Therefore,

$$\mu F_\rho = [\mu Pg - r, \mu Pg + r],$$

with $r := \operatorname{dist}(\mu,\Lambda)(\rho^2 - \|g^*\|^2)^{1/2}$. But, since $\rho^2 - \|g^*\|^2 \ge \|g\|^2 - \|Pg\|^2 = \operatorname{dist}(g,\hat{\Lambda})^2$, we could have obtained this bound from the earlier bound

$$|\mu g - \mu Pg| \le \operatorname{dist}(\mu,\Lambda)\operatorname{dist}(g,\hat{\Lambda}) .$$

Actual application of the optimal formula μPg requires that we construct $Pg = g^*$ which, in turn, requires that we know $\hat{\Lambda}$. This means that we must be able to <u>represent</u> the elements of Λ as points in X. Abstractly, this requires solution of an extremum problem: The representer $\hat{\lambda}$ for $\lambda \in X^*$ is the point in X at which the function

$$x \longmapsto \|x\|^2/2 - \lambda x$$

takes on its minimum. But, the only nontrivial concrete cases in which the optimal formula μPg for μg has been constructed have involved Hilbert spaces with reproducing kernel.

<u>Example: Optimal quadrature formulae for</u> $\mathbb{L}_2^{(k)}[a,b]$. Take

$$\mu : g \mapsto \int_a^b g(t)dt \ , \quad \lambda_i : g \mapsto g(t_i), \quad a \leq t_1 < \dots < t_n \leq b$$

and assume that, in addition to the numbers $(g(t_i))_1^n$, we have a bound on

$$v(g)^2 := \int_a^b (g^{(k)}(t))^2 \, dt \quad .$$

Then, the correct setup is the Hilbert space

$$X = \mathbb{L}_2^{(k)}[a,b] := \{x \in C^{(k-1)}[a,b] : x^{(k-1)} \text{ abs.cont.}, \ x^{(k)} \in \mathbb{L}_2[a,b]\}$$

with inner product

$$\langle x,y \rangle := \sum_{i=1}^{k} x(t_i)y(t_i) + \int_a^b x^{(k)}(t)y^{(k)}(t)dt \ .$$

This space has a reproducing kernel

$$K(s,t) := \sum_{i=1}^{k} \ell_i(s)\ell_i(t) + \int_a^b G(s,u)G(t,u)du \ ,$$

with

$$\ell_i(t) := \prod_{\substack{j=1 \\ j \neq i}}^{k} \frac{t-t_j}{t_i-t_j} \ , \quad G(\cdot,u) := (1-Q)(\cdot-u)_+^{k-1}/(k-1)!$$

$$Qx := \sum_{i=1}^{k} x(t_i)\ell_i \ .$$

This means that

$$x(t) = \langle x, K(\cdot,t) \rangle, \quad \text{for all } x \in X, \quad t \in [a,b].$$

Such a reproducing kernel allows one to find the representer $\hat{\lambda}$ for $\lambda \in X^*$ simply as $\hat{\lambda}(t) = \lambda K(\cdot,t)$. For our example, we have $\lambda_i : g \mapsto g(t_i)$, hence $\hat{\Lambda}$ is spanned by the sections $K(t_i,\cdot)$, $i=1,\dots,n$. The experienced eye recognizes these to be spline functions of order $2k$ with knots $t_1,\dots,t_n$. The center g^* of F_ρ therefore becomes the spline interpolant to g and the optimal formula μg^* is obtained by integrating this interpolant.

In summa , here is a very pretty theory, now almost twenty years old. It gives a mathematically rigorous way to go about evaluating integrals. It has generated many papers and continues to generate nice mathematics. Recently, for instance, it has become possible, through the theory of perfect splines, to derive optimal formulae in case the bound on g is in terms of the maximum norm of some derivative of g .

Why are these optimal formulae not used? There are several reasons for this. There is not just one formula, there are many different optimal formulae, corresponding to the many different ways in which one may estimate the size of the integrand g . In particular, all classical quadrature formulae are optimal in some sense. How is one to choose one? Further, if one uses such a formula in the way their mathematical designers intended them to be used, then one would have to obtain the required bound on g , possibly an expensive undertaking. For this, note that the calculation of (a bound for) $\int (g^{(k)}(t))^2 dt$ or $\max_t |g^{(k)}(t)|$ is a problem of the same complexity as the calculation of $\int g(t)dt$ itself.

But, most important of all, these formulae, when they are new at all, have not performed appreciatively differently from traditional formulae when used in the traditional way. True, these optimal formulae provide precise bounds on the number $\int g$ sought and, when coupled with interval arithmetic, they allow one to use calculations in an efficient way to prove theorems about the number $\int g$. But, the typical user of quadrature formulae does not see himself in the role of a theorem prover. There are usually too many aspects of his calculations which lack rigor and over which he has no control. He thinks of the numerical evaluation of an integral as a numerical experiment: Some quadrature rule is used to produce a guess for $\int g$, and this guess is then used in subsequent calculations as if it were correct. Only if contradictions arise in subsequent calculations, does he reexamine this guess and use, perhaps, a more sophisticated and more costly procedure for getting another guess . In extreme cases,he might even resort to an examination of the error term, typically given in terms of some derivative of the integrand,

$$\int_a^b g(t)dt = \text{rule}(h) + \text{const}\,(b-a)\,h^r\,g^{(r)}(\xi)$$

and so ascertain the order of magnitude of the error. But even then, it doesn't seem worth the effort to use an optimal rule to milk the bound on $g^{(r)}$ of all its information. If the rough bound is not good enough, it is usually cheaper to use a smaller step size h to get a guaranteed bound of prescribed size.

Automatic numerical quadrature. Faced with the bewildering collection of quadrature rules available, the typical user today will often choose a library subroutine which makes the choice for him, an automatic quadrature routine. The user provides means of evaluating the integrand g at any point of the interval of integration, and specifies a tolerance, and then leaves it to the routine to come up with an approximate answer. In terms of the scientific user described earlier, this makes good sense. To him such a routine represents a carefully conducted numerical experiment. All the obvious checks are supposedly applied, and, if the answer turns out to be wrong ultimately, the mistake must have been a subtle one and would not have been noted by him, either.

It is, of course, impossible for a mathematician as a mathematician to work on such a routine. The point underlying the Golomb-Weinberger paper is well taken: It is impossible to say anything about $\int g$ given only the values of g at finitely many points. Still, numerical analysis has a responsibility in this area.

The basic design of such a routine is quite simple. The domain D of integration is subdivided into an essentially disjoint union of cells (C_i). On each of these cells, a guess for $\int_{C_i} g$ along with an error estimate for that guess is derived. If the error estimate is no bigger than the error allowed for this cell, the guess is accepted and added to the total. Otherwise, the cell is further subdivided into smaller cells.

Both the guess and the estimate for its error are obtained by numerical experiment: Initial investigation of g on C_i leads to a model for its behavior, and guess and error estimate are based on the assumption that g conforms on C_i to that model.

One goes to a subdivision (C_i) of D at all because only for a sufficiently small cell C can standard models ("g is a polynomial of degree < 5", "g has a jump discontinuity", "$g(t)/(t - \alpha)^\beta$ is a polynomial of degree < 5", etc) be expected to reflect accurately actual relevant behavior. The goal is a suitably nonuniform subdivision, adapted to the particular integrand, each cell just small enough to allow a simple model of g to be sufficient and to have the corresponding error (estimate) just small enough.

It is possible [1] to bring such algorithms back into the realm of mathematics by assuming that the error estimate

$$\text{error}(C)$$

provided by the algorithm is, in fact, a bound on the error in the guess produced for $\int_C g$. Under such an assumption, it is then possible to prove that the work required to integrate certain badly behaved functions accurately is no bigger than the effort required for smooth integrands, the latter being taken as the ideal case.

[1] C. de Boor and J.R. Rice, An adaptive algorithm for multivariate approximation giving optimal convergence rates, MRC TSR 1773, 1977 .J.Approximation Theory, to appear.

[2] E.W. Cheney, Introduction to Approximation Theory, McGraw-Hill, New York, 1966

[3] M. Golomb and H.F. Weinberger, Optimal approximations and error bounds, in "On numerical approximation", R.E. Langer ed., Univ.Wisconsin Press, Madison, WI, 1959, 117-190

[4] J.R. Rice, The Approximation of Functions, Vols. I and II, Addison-Wesley, Reading, MA, 1964 and 1969 .

[5] T.J. Rivlin, An Introduction to the Approximation of Functions, Blaisdell Publ., Waltham, MA 1969

[6] A. Sard, Linear Approximation, Math. Surveys vol. 9, Amer. Math.Society, Providence, R.I., 1963

$$k_i = f(y_n + h \sum_{j<i} b_{ij}k_j) \ .$$

iving methods of this type which are of high order is extremely tedious, interested reader is referred to Butcher (1963, 1972) for the most cogent on. The maximum order of an r-stage method is still in doubt: there age methods of order 4, but one must go to a 6-stage method to achieve similarly there are 7-stage methods of order 6, but to achieve order must go to a 9-stage method. No more general results are known.

h methods are called explicit because each k_i can be explicitly calculated of quantities already known. Because of this, when applied to the oblem $y' = \lambda y$, one obtains

$$y_{n+1} = P(h\lambda)y_n \ ,$$

z) is a polynomial of order equal to the number of stages. Unfortunately, licit Runge-Kutta schemes, particularly those of high order, have rather ability regions. One can however, define general implicit Runge-Kutta

$$y_{n+1} = y_n + h(c_1k_1 + \ldots + c_nk_r)$$

$$k_i = f(y_n + h \sum_{j=1}^{r} b_{ij}k_j) \ .$$

equations for the $\{k_i\}$ are implicit, and form a rather formidable system near algebraic equations to solve, over each time step. They have es however: such an r-stage scheme can have order 2r, and, when applied odel problem $y' = \lambda y$, one obtains

$$y_{n+1} = R(h\lambda)y_n \ ,$$

z) is a rational function. The stability regions for such methods can

Proceedings of Symposia in Applied Mathematics
Volume 22
1978

Numerical Methods for the Solution of Ordinary Differential Equations

James M. Varah

University of British Columbia

These notes are divided in two parts: methods for initial value problems and boundary value problems. The emphasis will be on the practical mathematical elements of the current most efficient methods.

A. Initial Value Problems

These are problems of the form: find a (usually vector-valued) function $y(t)$ satisfying $y' = f(t,y)$, $y(a) = \alpha$, where f is a general nonlinear function of t and y. Existence of a unique solution over a bounded domain is assured provided f is Lipschitz in y. The methods we describe all find a discrete solution to this problem:

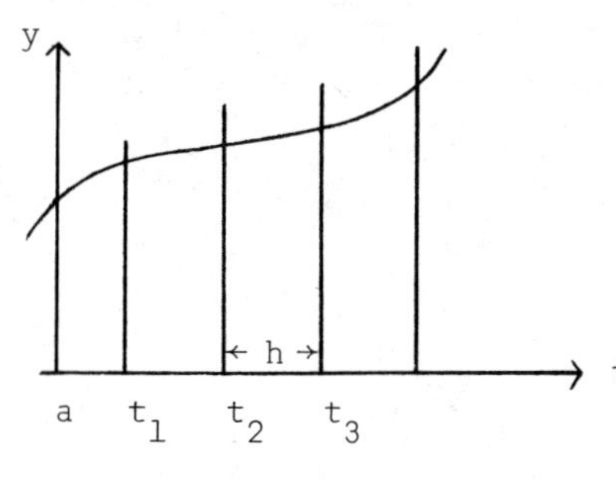

starting from $t_0 = a$, one finds an approximation y_j to $y(t_j)$, $j=1,2,\ldots$. For the mathematical exposition, we shall assume $t_j - t_{j-1} = h$, but in practice this stepsize changes during the course of the integration.

We say such a discrete method has order p if at any point t in the range of integration, the global error $|y(t) - y_{n(h)}| = O(h^p)$ as $h \to 0$ (here $n(h)$ denotes the discrete index where $t_{n(h)} = t$). Normally, if this global error is $O(h^p)$, the error made over one step, or local truncation error is $O(h^{p+1})$.

The simplest such method is the Euler method:

$$y_{j+1} = y_j + hf(t_j, y_j) \quad , \quad j=0,1,2,\dots \; .$$

Provided the solution y(t) is C_2, we can write

$$y(t_{j+1}) = y(t_j) + hf(t_j, y(t_j)) + \frac{h^2}{2} y''(\xi_j).$$

Thus if we take $y_j = y(t_j)$, this last term is the local truncation error; moreover using the Lipschitz constant and a bound for y"(t), it is easy to show that the global error is O(h). See for example Chapter 1 of Gear (1971).

Although the order of a method describes the convergence rate for a small enough stepsize, it does not give any information about how small h must be for this convergence to occur. Consider the scalar linear problem

$$y' = -100(y - \sin t) + \cos t \quad , \; y(0) = 0 \; .$$

This has the solution y(t) = sin t; however using Euler's method with different h, we obtain the following values at t = 1:

h	.01	.02	.03
y(1)	.84151	.83219	2.6×10^8

This behaviour is due to the presence of the decay factor (-100) in the equation, even though it does not appear in the solution. (Such equations, with large negative decay factors, are called <u>stiff</u> equations.)
Consider the even simpler model decay problem

$$y' = \lambda y \quad , \text{ with } \mathrm{Re}(\lambda) < 0 \; .$$

The Euler method applied to this gives

$$y_n = (1 + h\lambda) y_{n-1} = (1 + h\lambda)^n y_0 \; .$$

So although this converges (for fixed λ) as h
the numerical solution unless $|1 + h\lambda| < 1$. T
of h (for this method) is determined by the de
this concept to any step-by-step method: we d
S for the method as that part of the left half
the numerical solution to $y' = \lambda y$ decays. Of
the better a method can cope with stiff equati
describe, we shall mention the stability regic

The basic Euler method can be generalized
one-step (or Runge-Kutta) methods, and linear
each of these in turn.

1. <u>Runge-Kutta Methods</u>

These methods keep the characteristic of
mation; however they "bootstrap" their way acr
can achieve higher order of accuracy than the
scheme is a two-stage explicit scheme:

$$y_{n+1} = y_n + h(c_1 k_1 +$$

where

$$k_1 = f(y_n) \quad , \quad k_2 = f(y$$

(For this discussion on Runge-Kutta methods, w
of f on t in the hope of simplifying the notat
c_1, c_2, and b_1, the method can be made second-c
scheme has the form

$$y_{n+1} = y_n + h(c_1 k_1 + \dots$$

where

and tl
discus
are 4-
order
7, or
S
in ter
model

where
most e
small
scheme

where

Now the
of non
advanta
to the

where R

be much larger than for explicit methods, even unbounded; in fact for the r-stage scheme of order 2r, R(z) is the r-th diagonal Padé approximation to e^z, which has $|R(z)| < 1$ for $Re(z) < 0$. Thus the stability region is the entire left half-plane.

However the necessity of solving nonlinear algebraic systems has precluded the general use of such schemes. In practice, only the explicit Runge-Kutta methods are used to any great extent, and can be competitive with multistep schemes on non-stiff problems. In particular the Runge-Kutta-Fehlberg formulas, which also provide error estimates, are to be recommended. Notice that for each step of an r-stage scheme, r evaluations of f(y) are required, which is more than what is required for a multistep scheme of the same order. Thus the Runge-Kutta formulas are useful only if the functions being integrated are not terribly complicated. For more information on these practical computational aspects of Runge-Kutta methods, we refer to Shampine et al (1976) and Enright and Hull (1976).

2. Multistep Methods

These methods generalize Euler's method in another way: they use more previous values of y_j and $f(t_j,y_j)$. Thus a 2-step explicit method would look like

$$y_{n+1} = a_1y_n + a_2y_{n-1} + b_1hf(t_n,y_n) + b_2hf(t_{n-1},y_{n-1}),$$

and the constants can be chosen so the method is higher order. The general r-step method can be written as

$$\sum_{i=0}^{r} \alpha_i y_{n+i} = h \sum_{i=0}^{r} \beta_i f(t_{n+i},y_{n+i}) \quad , \quad n=0,1,2,\ldots .$$

Notice that this defines y_{n+r} explicitly if $\beta_r = 0$ and implicitly if $\beta_r \neq 0$; thus we have explicit and implicit multistep methods.

It may appear that there are enough coefficients to make such an r-step formula of order 2r; however there are extraneous solutions to the multistep scheme to worry about. Consider such a general scheme applied to the model problem $y' = \lambda y$:

$$\sum_0^r \alpha_i y_{n+i} = h\lambda \sum_0^r \beta_i y_{n+i} \quad , \quad n=0,1,2,\ldots .$$

This is a linear homogeneous difference equation with constant coefficients, and has the general solution

$$y_n = \sum_{i=1}^r c_i (\xi_i(h\lambda))^n ,$$

where $\{\xi_i(h\lambda)\}$ are the roots of the characteristic polynomial

$$\rho(\xi) \equiv \sum_0^r \alpha_i \xi^i = h\lambda(\sum_0^r \beta_i \xi^i) \equiv h\lambda\sigma(\xi) .$$

Thus the region of absolute stability S is that part of the left half of the complex z-plane where all roots $\xi_i(z)$ of $\rho(\xi) - z\sigma(\xi) = 0$ have $|\xi_i| < 1$. In particular, $z = 0$ should belong to S (or more precisely to the boundary δS since one root $\xi_1(0) = 1$); if not, then the numerical solution to $y' = 0$ will explode as $h \to 0$. Such methods (where $0 \in \delta S$) are called stable; Dahlquist (1956) showed that this was the crucial property of a multistep method: stable methods of order p actually converge at this rate as $h \to 0$. At the same time, Dahlquist also showed that stable r-step methods had maximum order $r + 1$. Moreover such methods actually exist: the implicit Adams methods of the form

$$y_{n+1} = y_n + h \sum_{i=0}^r b_i f(t_{n-i+1}, y_{n-i+1}) .$$

The implicit Adams methods are widely used to solve non-stiff equations; the implicit equation is solved using the above formula as a direct iteration for y_{n+1}, using as an initial approximation the result of the corresponding explicit Adams method

$$y_{n+1} = y_n + h \sum_{i=0}^{r-1} b_i^* f(t_{n-i}, y_{n-i}) .$$

This is called the predictor-corrector technique, and normally only one or two iterations are required.

Unfortunately these Adams methods again have rather small stability regions, typically of the form

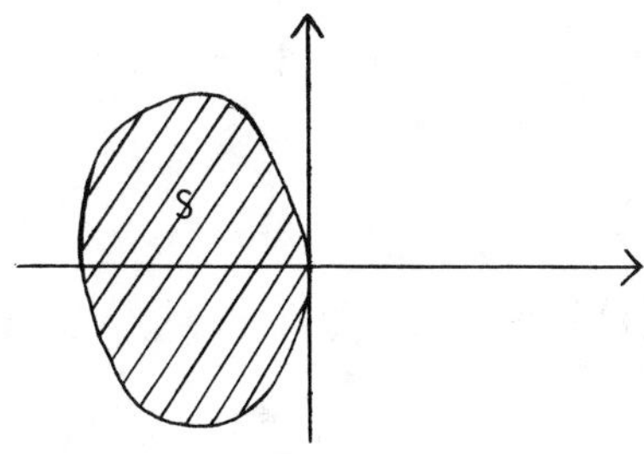

and thus they are not useful for stiff equations. For this purpose it would be of interest to derive methods having unbounded S:

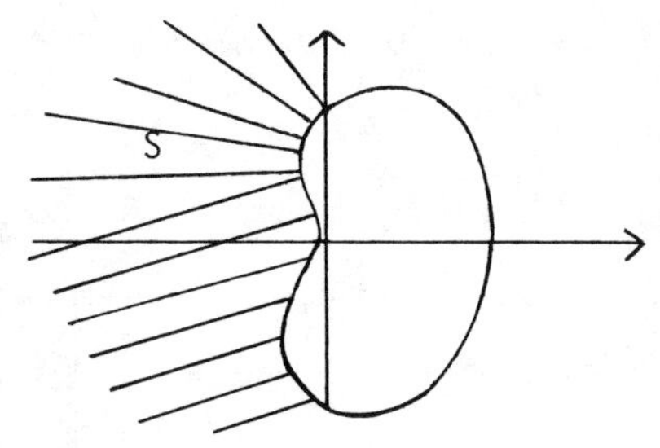

Clearly such a situation exists when $\infty \in S$, that is when the polynomial $\sigma(\xi)$ has all the r roots inside the unit circle. Notice that this immediately excludes explicit schemes: these have $\sigma(\xi)$ of degree (r-1) and thus one root $|\xi_i(z)| \to \infty$ as $z \to \infty$. In the same spirit as the Dahlquist result quoted above, one can show that $\infty \in S$ only for r-step methods of order r or less. Again, such methods actually exist; an example is the backward differentiation formula

$$y_{n+1} = \sum_{i=0}^{r-1} a_i y_{n-i} + hb_0 f(t_{n+1}, y_{n+1}) .$$

These methods have $\sigma(\xi) = \xi^r$, and hence have the additional feature that as $\text{Re}(h\lambda) \to -\infty$, they give more decay since all $\xi_i(h\lambda) \to 0$. These methods are, however, stable only for $r \leq 6$; a widely used software package for stiff equations using this scheme has been developed by Gear (1971). One can in fact extend these methods to stable schemes of higher order by increasing r and demanding that the order be less than r, then using the resulting freedom in the coefficients to keep stability and increase S as much as possible. This has led to 10-step methods of order 7 and 12-step methods of order 8 (see Varah (1978)).

We do not intend to go into a discussion of what makes a good numerical package for integrating a general system of equations. The best packages are based on the schemes we have mentioned, but their success depends to a large extent on how the stepsize (and order) are changed during the course of the integration, and how the errors are estimated. For such discussions and comparisons of methods, we refer to Shampine et al (1976), Enright and Hull (1976), and Enright et al (1975).

Instead, let us return to a question glossed over in our discussion of stability regions. Suppose a method has a reasonably large stability region S, so that for $h\lambda \in S$, the method decays numerically when applied to the problem $y' = \lambda y$. Does this imply that the scheme will behave "appropriately" on a more general equation, for example on the linear system

$$y' = A(t)y + F(t) .$$

In a recent paper, Kreiss (1977) has shown that the answer is no in general; for some systems, the behaviour of the solution is not determined completely by the eigenvalues of $A(t)$. For example, the eigenvalues may all be negative, yet the solution may not decay. In such a case, it is clear that the analysis using the stability region S is not adequate to say whether a method is accurately reflecting the nature of the solution. However, Kreiss also shows that if the linear system is in a certain form (negative dominant form), then the nature of the solution is accurately represented by the eigenvalues; moreover if a method has a reasonable stability region S, then applied to such a problem, the numerical solution it generates does accurately reflect the nature of the true solution.

B. Boundary Value Problems

The kinds of boundary value problems which arise, and the existence and uniqueness of solution, are much more complex questions than for initial-value problems. We shall restrict ourselves to two-point boundary value problems, for which the typical formulation may be

(i) $Ly(t) = g(t)$ for $a \le t \le b$, with L some linear differential operator of order n; associated with this are n boundary conditions involving the scalar function $y(t)$ and possibly its first $n-1$ derivatives at $t = a$ and $t = b$.

or

(ii) $y' = K(t)y + f(t)$ for $a \le t \le b$, where $y(t)$ is an n-vector; again there are n boundary conditions involving $y(a)$, $y(b)$.

or

(iii) $y' = f(t,y)$ for $a \le t \le b$, where f is some general nonlinear vector function of the n-vector $y(t)$; again there are n (possibly nonlinear) boundary conditions involving $y(a), y(b)$.

Of course (i) ⊂ (ii) ⊂ (iii), but the method used to solve a particular problem may depend on which form the problem has. We now describe the basic methods.

1. Shooting Methods

Assume the problem is given in form (iii); then the problem can be converted to an initial value problem (say starting at $t = a$) by introducing extra boundary conditions at $t = a$. Now the problem is integrated for $a \leq t \leq b$ using some initial value method, and the values obtained at $t = b$ checked against the given boundary conditions at $t = b$. One then adjusts the extra boundary conditions until the values obtained at $t = b$ agree with the given boundary conditions. The order of accuracy attained is that of the initial value method used.

Unfortunately, the behaviour of the solution to the initial value problem may be very different from that of the boundary value problem; one may obtain exponentially increasing solutions which blow up before $t = b$ is reached, and the solution may be very sensitive to the coefficients involved in the extra boundary conditions. To rectify this, one often "shoots" across $[a,b]$ using several smaller intervals, matching the solutions by continuity conditions (multiple shooting). For more discussion and results, we refer to Keller (1976) and Roberts and Shipman (1972).

Despite these drawbacks, multiple shooting has proven to be a very effective method; for discussion of a particularly good implementation and numerical results, we refer to Scott and Watts (1977).

2. Finite Difference Methods

Ten years ago, a typical finite difference scheme (applied to form (i)) was to put some equally-spaced mesh $\{t_j = a + jh,\ j=0,\ldots,n\}$ on $[a,b]$, approximate the derivatives appearing in the differential equation by differences, and apply the resulting difference equation at each mesh point; including the boundary conditions as well, this gave a system of linear equations $A\underline{u} = \underline{g}$ to solve for the approximations $\{u_j\}$ to $\{y(t_j)\}$. If p-th order approximations were used for the derivatives, and if the matrix A was well-conditioned, one could show the errors $|u_j - y(t_j)| = O(h^p)$.

More recently, finite difference schemes have become more flexible and versatile; we mention specifically the "box scheme" of Keller (1968). Applied to an equation in form (ii), the scheme is

$$u_{j+1} - u_j = h_j K(t_{j+\frac{1}{2}}) \left(\frac{u_{j+1} + u_j}{2} \right) + h_j f(t_{j+\frac{1}{2}})$$

(Here $t_{j+\frac{1}{2}} = (t_j + t_{j+1})/2$.)
Since the scheme involves only two successive meshpoints, the mesh $\{t_j\}$ can be arbitrarily spaced, and yet still maintain $O(h^2)$ error as $h = \max |h_j| \to 0$. We again are led to a linear system to solve, with the matrix A in block-bidiagonal form.

Although one can derive higher-order difference schemes in a similar vein, one can achieve higher-order accuracy more simply by applying extrapolation or deferred correction techniques to the above scheme. The basis for this is the asymptotic expansion of the error in this scheme (or any scheme you wish to start with). Here we have

$$e(t_j) \equiv u_j - y(t_j) = h^2 e_1(t_j) + h^4 e_2(t_j) + \ldots + h^{2m} e_m(t_j) + O(h^{2m+2}) .$$

One first solves the discrete problem over the given mesh $\{t_j\}$ giving $u_j^{(0)}$. For extrapolation, the mesh is refined by introducing new points halfway between the $\{t_j\}$; this new discrete system is solved, obtaining new approximations at all the points of this new mesh, which includes all the points of the original mesh $\{t_j\}$ (call these $u_j^{(1)}$). Now if we take the right linear combination of $u_j^{(0)}$ and $u_j^{(1)}$, we remove the term $e_1(t_j)$ from the error expansion; in fact we have

$$u_j^{(1)} + \frac{1}{3}(u_j^{(1)} - u_j^{(0)}) = y(t_j) + O(h^4).$$

This extrapolation process can be repeated, by refining the mesh further and using the values obtained to remove further terms in the error expansion. Notice however that the size of the linear system increases with the order of extrapolation; the deferred correction approach removes this difficulty by solving successively (on the same mesh) for approximations to the leading terms $\{e_i(t_j)\}$ in the error expansion. In fact, the error term $e(t_j)$ satisfies $A\underline{e} = \underline{\tau}$ where $\tau(t_j)$ is the local truncation error. Thus one solves for the error terms using the same matrix A and a succession of higher order approximations to the truncation error. See Pereyra (1968) for more details.

This basic scheme can also be applied to nonlinear equations (Keller (1974)): for the nonlinear system $y' = f(t,y)$ the box scheme is

$$u_{j+1} - u_j = h_j f\left(t_{j+\frac{1}{2}}, \frac{u_{j+1} + u_j}{2}\right)$$

This gives a system of nonlinear algebraic equations to solve, $G(\underline{u}) = 0$. Results on existence and uniqueness of solutions to such equations are very difficult to obtain, but in practice one iterates towards a solution by Newton's method: start from some initial guess $\underline{u}^{(0)}$ and solve the sequence of linear systems

$$J(\underline{u}^{(k)})(\underline{u}^{(k+1)} - \underline{u}^{(k)}) = -G(\underline{u}^{(k)}) \quad , \quad k=0,1,2,\ldots ,$$

where $J_{ij}(u) = \frac{\partial G_i(u)}{\partial u_j}$ = Jacobian of the system.

As might be expected, the above processes can be automated, and codes are available which attempt to solve any given nonlinear system of boundary value problems (form (iii)) to a given accuracy using the box scheme plus deferred correction (see Lentini and Pereyra (1977)).

3. Projection Methods

These are the most recent methods, dating only from the last ten years. Consider the problem given in form (i): $Ly = g$. Then one asks for an approximate solution

$$u^{(n)}(t) = \sum_1^n c_i \phi_i(t)$$

where the $\{\phi_i(t)\}$ are given functions. Thus the approximate solution $u^{(n)}(t) \in S_n$, the discrete space spanned by $\{\phi_1(t),\ldots,\phi_n(t)\}$. The coefficients $\{c_i\}$ are to be chosen so that $u^{(n)}(t) \cong y(t)$. This can be done in different ways:

(a) $\int_a^b (Lu^{(n)} - g)\phi_j dt = 0$, $j=1,\ldots,n$ (Galerkin's method)

(b) $Lu^{(n)}(t_j) = g(t_j)$, $j=1,\ldots,n$ (collocation)

(c) $\int_a^b (Lu^{(n)} - g)(L\phi_j)dt = 0$, $j=1,\ldots,n$ (least squares) .

Each of these leads to a linear system $A\underline{c} = \underline{b}$, and the resulting approximations normally turn out to have about the same accuracy, if done properly. However the collocation method has clear computational advantages: no integrals are involved.

In particular, if no specific choice is indicated for the $\{\phi_i(t)\}$ one can use piecewise polynomials over a given mesh $\{t_j\}$, with certain continuity assumed at the mesh points, resulting in spline approximations. Moreover, one can choose bases for the piecewise polynomial spaces so that each element $\phi_i(t)$ has support over only a few subintervals: for example

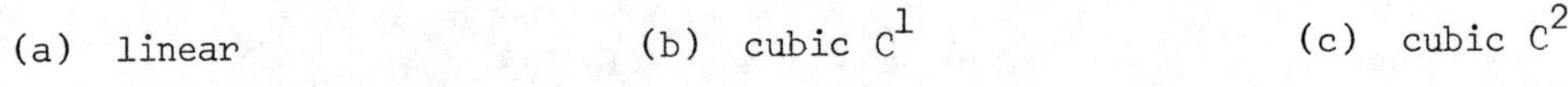

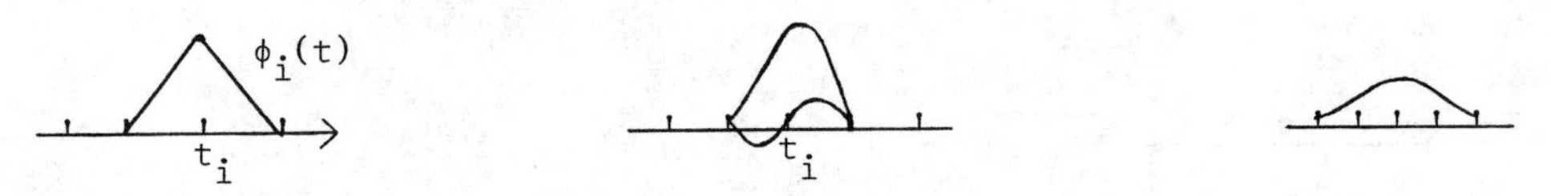

This means that the matrix A will be banded, and it also forces the matrix to be well conditioned.

The accuracy possible with such schemes is naturally connected to the approximation properties of the underlying discrete space S_n. For piecewise polynomial spaces over a mesh of basic size h, it is possible to achieve $O(h^{2m})$ error for piecewise polynomials of degree $(2m - 1)$, using Galerkin, collocation, or least squares. For collocation, it is also necessary to use low order continuity at the mesh points, and to collocate at specific (Gaussian) points in each subinterval. (Thus for example collocating with C^1 cubics at Gaussian points gives $O(h^4)$ error.) This was first pointed out by de Boor and Swartz (1973), and has made the computationally more attractive collocation schemes the most useful of the projection methods. In certain cases, even higher order convergence, at the mesh points only, can be obtained; this is known as supercovergence.

Speaking from a practical point of view, we have personally found these collocation schemes to be extremely useful; they have worked very well on a variety of different problems. No general purpose collocation code is yet available, but one is being developed by Ascher et al (1977): preliminary results show it is very competitive with the best codes for finite-difference and shooting methods.

Finally, we do not want to leave the reader with the impression that all these boundary value problems are easily solved by the methods we have outlined. There are many problems which require special attention: problems with singularities of the endpoints for example - these problems have solutions with singular function expansions which should be included in the numerical solution. Also, there are singular perturbation problems; for example the linear problem

$$\varepsilon y'' + a(t)y' + b(t)y = g(t) \quad , \quad a \le t \le b$$

where ε is a small parameter. Solutions to these problems can have boundary layers at the endpoints, and as well interval boundary layers near the zeros of $a(t)$ (turning points). Good solutions to these problems cannot be obtained without careful attention to these boundary layers. And these are nonlinear problems with multiple solutions: for example bifurcation problems of the form

$$y'' = f(y,t,\lambda)$$

where λ is a parameter. Problems like this can have any number of solutions depending on the value of λ; the corresponding behaviour of the discrete non-linear system (obtained from a finite difference method for example) is not well understood.

References

U. Ascher, J. Christiansen, and R. Russell (1977), A collocation solver for mixed order systems of boundary value problems. Technical Report 77-13, Computer Science Dept., University of British Columbia, Vancouver, Canada.

J.C. Butcher (1963), Coefficients for the study of Runge-Kutta integration processes. J. Aust. Math-Soc. 3, pg. 185-201.

J.C. Butcher (1972), An algebraic theory of integration methods. Math. Comp. 26, pg. 79-106.

G.G. Dahlquist (1956), Numerical integration of ordinary differential equations. Math. Scand. 4, pg. 33-50.

G.G. Dahlquist (1963), A special stability problem for linear multistep methods. BIT 3, pg. 27-43.

C. de Boor and B. Swartz (1973), Collocation at Gaussian points. SIAM J. Num. Anal. 10, pg. 582-606.

W. Enright, T. Hull, and B. Lindberg (1975), Comparing numerical methods for stiff systems of ordinary differential equations. BIT 15, 10-48.

W. Enright and T. Hull (1976), Test results on initial value methods for non-stiff ordinary differential equations. SIAM J. Num. Anal. 13, pg. 944-161.

C.W. Gear (1971), Numerical Initial Value Problems in Ordinary Differential Equations. Prentice-Hall, Englewood Cliffs, N.J.

H.B. Keller (1968), Accurate difference methods for linear ordinary differential equations subject to linear constraints. SIAM J. Num. Anal. 6, pg. 8-30.

H.B. Keller (1974), Accurate difference methods for nonlinear two-point boundary value problems. SIAM J. Num. Anal. 11, pg. 305-320.

H.B. Keller (1976), Numerical Solution of Two-Point Boundary Value Problems. SIAM Publication - CBMS Regional Conference Series in Applied Mathematics.

H.O. Kreiss (1977), Difference methods for stiff ordinary differential equations. SIAM J. Num. Anal. (to appear).

J.D. Lambert (1973), Computational Methods in Ordinary Differential Equations. John Wiley & Sons, New York.

M. Lentini and V. Pereyra (1977), An adaptive finite difference solver for nonlinear two-point boundary problems with mild boundary layers. SIAM J. Num. Anal. 14, pg. 91-111.

V. Pereyra (1968), Iterated deferred corrections for nonlinear boundary value problems. Num. Math. 11, pg. 111-125.

S. M. Roberts and J.S. Shipman (1972), Two-Point Boundary Value Problems: Shooting Methods. American Elsevier, New York.

M.R. Scott and H.A. Watts (1977), Computational solutions of linear two-point boundary value problems via orthonormalization. SIAM J. Num. Anal. 14, pg. 40-70.

L.F. Shampine, H.A. Watts, and S.M. Davenport (1976), Solving non-stiff ordinary differential equations - the state of the art. SIAM Review 18, pg. 376-411.

H.J. Stetter (1975), Analysis of Discretization Methods of Ordinary Differential Equations. Springer-Verlag, New York.

J.M. Varah (1978), Stiffly stable linear multistep methods of extended order. SIAM J. Num. Anal. (to appear).

Proceedings of Symposia in Applied Mathematics
Volume 22
1978

METHODS FOR TIME DEPENDENT PARTIAL DIFFERENTIAL EQUATIONS

JOSEPH OLIGER
Computer Science Department
Stanford University
Stanford, CA 94305

0. Introduction

I am going to discuss approximate methods for time dependent partial differential equations and some properties of the resulting algorithms which hopefully produce the desired approximation. I will begin with the oldest general technique, difference methods, which still dominates the consumer's market. In the first section we will look at the convergence theory for difference methods, sketch two lines of stability analysis--Fourier techniques and energy techniques, and look at an application of the latter to an approximation of the shallow-water equations. In the second section we will consider the derivation and form of other types of approximations--finite element and collocation methods. We will make a brief comparison of the analysis of these methods with that of difference methods and discuss some of the areas of application where these methods are used. In the third section we consider the usefulness and impact of "fast algorithms" for subproblems on the general development of methods for time dependent problems--the fast Fourier transform is one such example. In the fourth section we will look at the influence that machine design can have on whether or not a given method is useful in a given computational environment. In the fifth section we will briefly discuss some recent results which are fruitful products of the application of techniques from other areas of pure and applied mathematics to numerical methods.

1. Difference Methods

We will limit our discussion to approximations of the Cauchy problem for linear systems of equations and then indicate some extensions and where they are treated. We refer to the book of Richtmyer and Morton [13] and the review paper of Thomée [17] for details. Consider the Cauchy problem for the linear system

$$\begin{aligned} u_t &= P(x,t,\partial_x)u, \quad x \in \mathbb{R}^s, \quad t \geq 0 \\ u(x,0) &= f(x), \quad x \in \mathbb{R}^s \end{aligned} \tag{1.1}$$

where $u = (u_1,\ldots,u_n)'$ and $f = (f_1,\ldots,f_n)'$. P is a polynomial in $\partial/\partial_{x_1},\ldots,\partial/\partial_{x_s}$ with matrix coefficients which have sufficiently many bounded derivatives. We assume this problem to be well-posed in $L_2(\mathbb{R}^s)$, i.e., there are constants $K,\alpha > 0$ such that

$$\|u(x,t)\| \leq Ke^{\alpha t}\|f(x)\|$$

for $f \in L_2$. Now approximate (1.1) by the two level approximation

$$\begin{aligned} (I+Q_0)v(x,t+k) &= (I+Q_1)v(x,t), \quad x \in \mathbb{R}^s, \quad t = 0,k,2k,\ldots \\ v(x,0) &= f(x), \quad x \in \mathbb{R}^s . \end{aligned} \tag{1.2}$$

This is no real loss of generality since multilevel methods can be written as two level methods by introducing new variables. The difference operators Q_j are polynomials with terms of the form

$$A(x,t,h)E^{\nu} = A(x,t,h)E_1^{\nu_1} \cdots E_s^{\nu_s}$$

where $\nu = (\nu_1,\ldots,\nu_s)$, the ν_j are integers, and the E_j are shift operators

$$E_j v(x) = v(x+he_j) .$$

e_j is the unit vector in the positive x_j-direction and $h,k > 0$ are discretization parameters. We assume that $\|(I+Q_0)^{-1}\| \leq M$ in L_2 and that the matrix coefficients of (1.2) are at least Lipschitz continuous. We now define the notions of accuracy and stability.

Definition 1.1. The approximation (1.2) is said to be accurate of order (q_1,q_2) for a particular solution u of (1.1) if there is a function $C(t)$, bounded on every finite interval $[0,T]$, such that

$$\|(I+Q_0)u(x,t+k) - (I+Q_1)u(x,t)\| \leq kC(t)(h^{q_1}+k^{q_2}) \tag{1.3}$$

for sufficiently small k and h. If $q = q_1 = q_2$ we will simply say that the method is accurate of order q. If $\min(q_1,q_2) \geq 1$ we will say that the method is consistent.

Definition 1.2. The approximation (1.2) is said to be strongly stable for sequences $h_\nu \to 0$, $k_\nu \to 0$ if there are constants α_s, K_s independent of h_ν, k_ν such that the solutions of (1.2) satisfy

$$\|v(x,t)\| \leq K_s e^{\alpha_s t} \|f(x)\| . \tag{1.4}$$

The utility of these properties is demonstrated by the following theorem which is a version of one direction of the well known Lax-Richtmyer theorem. An important feature of this result is that it provides an error estimate for discrete values of k and h.

Theorem 1.1. Let u and v be the solutions of (1.1) and (1.2), respectively. Assume that the estimates (1.3) and (1.4) are valid for fixed k and h, then

$$\|u(x,t)-v(x,t)\| \leq K_s \max_{0\leq\tau\leq t} \|(I+Q_0)^{-1}\| \max_{0\leq\tau\leq t} \|C(\tau)\| \cdot (h^{q_1}+k^{q_2})\ell(\alpha_s,t) \tag{1.5}$$

where

$$\ell(\alpha_s,t) = \begin{cases} e^{\alpha_s k}(e^{\alpha_s t}-1)/\alpha_s & \text{if} \quad \alpha_s \neq 0 \\ t & \text{if} \quad \alpha_s = 0 \ . \end{cases}$$

The usefulness of Theorem 1.1 depends upon how easily we can establish its hypotheses. The accuracy properties are easily established for smooth solutions using Taylor series expansions. For example, consider the symmetric hyperbolic system

$$\begin{aligned} u_t &= Au_x \quad , \quad x \in \mathbb{R} \ , \quad t \geq 0 \\ u(x,0) &= f(x) \ , \quad x \in \mathbb{R} \end{aligned} \tag{1.6}$$

where A is a real symmetric matrix. If we approximate (1.6) by

$$\begin{aligned} v(x,t+k) &= (I+kAD_0 + (k^2/2)A^2D_+D_-)v(x,t) \\ v(x,0) &= f(x) \end{aligned} \tag{1.7}$$

where

$$D_0v(x,t) = (v(x+h,t) - v(x-h,t))/2h$$

$$D_+D_-v(x,t) = (v(x+h,t) - 2v(x,t) + v(x-h,t))/h^2$$

then

$$\|(I+Q_0)u(x,t+k) - (I+Q_1)u(x,t)\| = O(k^3+kh^2)$$

since

$$D_0u(x,t) = u_x(x,t) + O(h^2)$$

$$D_+D_-u(x,t) = u_{xx}(x,t) + O(h^2)$$

and

$$u(x,t+k) = u(x,t) + ku_t(x,t) + (k^2/2)u_{tt}(x,t) + O(k^3)$$

$$= u(x,t) + kAu_x(x,t) + (k^2/2)A^2u_{xx}(x,t) + O(k^3)$$

$$= u(x,t) + kAD_0u(x,t) + (k^2/2)A^2D_+D_-u(x,t) + O(kh^2+k^3)$$

follow from Taylor expansions of the solution about the point (x,t).

Stability is generally more difficult to establish. There are two methods of approach that are commonly used--Fourier techniques and energy methods. We will begin with the Fourier techniques.

Consider the system (1.1) with constant coefficients and approximations (1.2) thereof whose coefficients $Q_j = Q_j(k,h)$ are functions of only h and k. We define the symbol of the method (1.2) to be

$$\hat{S} = (I+\hat{Q}_0)^{-1}(I+\hat{Q}_1) \tag{1.8}$$

where

$$\hat{Q}_j(\omega,h,k) = e^{-i\langle\omega,x\rangle}Q_je^{i\langle\omega,x\rangle}\ , \quad j = 0,1,$$

$\omega = (\omega_1,\dots,\omega_s)$ is a real vector and $\langle\cdot,\cdot\rangle$ denotes inner product. It then easily follows using the Fourier transform and Parseval's relation that (1.2) is strongly stable if and only if

$$|\hat{S}^{t/k}| \leq K_se^{\alpha_st} \tag{1.9}$$

for all ω, h_ν and k_ν. The inequality (1.9) reduces the stability problem to an algebraic question: When is the family of matrices $\hat{S}(\omega,h_\nu,k_\nu)$ power bounded? If $\hat{S}$ is a normal matrix, $\hat{S}\hat{S}^* = \hat{S}^*\hat{S}$, then (1.9) follows if

$$|\varkappa(\hat{S})| \leq e^{\alpha_sk} \tag{1.10}$$

where $\varkappa(\hat{S})$ denotes the eigenvalues of $\hat{S}$. The condition (1.10) is always necessary for stability and is called the von Neumann necessary condition. General sufficient conditions are given by, e.g., the Kreiss matrix theorem (see [13]), but they are quite complicated and we will not pursue them here. There is an important class of methods where (1.9), and hence stability, only depends on properties of the eigenvalues.

Definition 1.3. The approximation (1.2) is dissipative of integral order $2r > 0$ if there are constants α_s and $\delta > 0$ such that

$$|\varkappa(\hat{S})| \leq e^{\alpha_s k}(1-\delta|\xi|^{2r}) , \quad |\xi| \leq \pi , \tag{1.11}$$

where $\xi = h\omega$.

The Lax-Wendroff method (1.7) is easily seen to be dissipative of order 4 if $(k/h)\max_j|\mu_j| < 1$ where the μ_j are the eigenvalues of the matrix A.

The following two theorems are important justifications of the concept of dissipativity.

Theorem 1.2. Let (1.2) approximate a symmetric hyperbolic system and assume that the coefficients of (1.2) are symmetric matrices. If (1.2) is accurate of order $2r-2$ or $2r-1$ and dissipative of order $2r$, then it is strongly stable.

Theorem 1.3. Let (1.2) be a consistent approximation to a parabolic system of order m, and assume that $k/h^m \leq$ const. If (1.2) is dissipative (of any order), then it is strongly stable.

We now turn to (1.1) with variable coefficients and its approximation (1.2). It is natural to consider the family of methods with constant coefficients of the form (1.2) which one obtains by "freezing" the coefficients of (1.2) at each point (x,t) in the domain. These are all problems we know how to attack. The question is: Does the stability of all these related constant coefficient problems imply the stability of the variable coefficient approximation? The answer to this question cannot be completely answered at present but such freezing results cannot be expected to hold except for approximations to hyperbolic and parabolic equations which are stable to lower order perturbations. Some continuity assumptions are also necessary to hold all of the pieces together. In the case of dissipative approximations, and under our assumption of Lipschitz continuity of the coefficients, we have the following results at our disposal, among others.

Definition 1.4. We will say that (1.2) is dissipative of order $2r$ if there is a $\delta > 0$ such that (1.11) is uniformly valid for all of the frozen systems.

Theorem 1.4. If (1.2) approximates a symmetric hyperbolic system and has symmetric coefficients, is accurate of order $2r-1$ and dissipative of order $2r$, then it is strongly stable.

In some circumstances the order of accuracy can be reduced to $2r-2$ in the above theorem (see [13]).

Theorem 1.5. Let (1.2) be a consistent approximation to a parabolic system of order m and assume that $k/h^m \leq$ const. If the approximation is dissipative (of any order), then it is strongly stable.

Since it is natural that approximations of parabolic equations be dissipative, Theorem 1.5 is quite satisfactory. An important extension of freezing arguments via pseudo-differential operators for non-dissipative approximations of hyperbolic equations has been carried out by Lax and Nirenberg [8]. These results require that the coefficients be smoother, which is not surprising since non-dissipative approximations do not have a built in smoothing property as do dissipative approximations.

We close this discussion of Fourier techniques with the remark that there is a similar, less complete, but viable theory for approximations of the initial boundary value problem for hyperbolic and parabolic equations which is based on Fourier-Laplace transform techniques following work of Kreiss. This work is discussed in, e.g., papers of Gustafsson, Kreiss and Sundström [5], Gustafsson [4], and Varah [18].

The energy method is less devious--it is a frontal attempt to establish the desired estimates using discrete analogs of integration by parts. It is more general in the sense that it can be used successfully to analyze approximations to problems which have variable coefficients and are not hyperbolic or parabolic. But it is less algorithmic and its successful application is often dependent on the cleverness of the investigator and the size of his bag of tricks. However, the results are usually more satisfactory when successfully obtained since they are less removed from the problem and often sharper as a consequence. We will now present a general result of this type which is often useful.

Consider the three level approximation

$$(I-kQ_1)v(x,t+k) = (I+kQ_1)v(x,t-k) + 2kQ_0v(x,t) \tag{1.12}$$

of the Cauchy problem (1.1), where the Q_j's are as previously defined. The following theorem holds [20] for the variable coefficient problem.

<u>Theorem 1.6</u>. The approximation (1.12) is strongly stable if there is a constant $\eta > 0$ such that

$$\begin{aligned} (v,Q_0w) &= -(Q_0v,w) , \\ k\|Q_0\| &< 1 - \eta . \end{aligned} \tag{1.13}$$

and

$$\mathrm{Real}(v,Q_1v) \leq 0$$

for all $v,w \in L_2$.

An Application

We will now look at an application of theorem 1.6 to an approximation of the shallow water equations. This is a system of equations which is typical of many problems in meteorology and oceanography. These equations can be written in the form

$$\begin{pmatrix} u \\ v \\ \varphi \end{pmatrix}_t + \begin{pmatrix} u & 0 & 1 \\ 0 & u & 0 \\ \varphi & 0 & u \end{pmatrix}\begin{pmatrix} u \\ v \\ \varphi \end{pmatrix}_x + \begin{pmatrix} v & 0 & 0 \\ 0 & v & 1 \\ 0 & \varphi & v \end{pmatrix}\begin{pmatrix} u \\ v \\ \varphi \end{pmatrix}_y = 0 \tag{1.14}$$

For global meteorological calculations the natural setting is a periodic boundary problem in x and y and it is more natural to write the equations in spherical coordinates, but we will refrain from doing so to keep the notation as simple as possible. The periodic boundary problem is technically equivalent to the Cauchy problem--we use Fourier series instead of integrals. With this slight modification of tools everything else follows as before. In this application $\varphi > 0$ and $|\varphi| \gg |u| + |v|$. We can make a change of variable

$$\begin{pmatrix} \varphi^{1/2} & 0 & 0 \\ 0 & \varphi^{1/2} & 0 \\ 0 & 0 & 1 \end{pmatrix}\begin{pmatrix} u \\ v \\ \varphi \end{pmatrix} = \begin{pmatrix} \tilde{u} \\ \tilde{v} \\ \tilde{\varphi} \end{pmatrix} \equiv w$$

which results in the symmetric system

$$w_t + \begin{pmatrix} u & 0 & \varphi^{1/2} \\ 0 & u & 0 \\ \varphi^{1/2} & 0 & u \end{pmatrix} w_x + \begin{pmatrix} v & 0 & 0 \\ 0 & v & \varphi^{1/2} \\ 0 & \varphi^{1/2} & v \end{pmatrix} w_y = 0 \tag{1.15}$$

We begin by looking at the frozen constant coefficient problem

$$w_t + \begin{pmatrix} u_0 & 0 & \varphi_0^{1/2} \\ 0 & u_0 & 0 \\ \varphi_0^{1/2} & 0 & u_0 \end{pmatrix} w_x + \begin{pmatrix} v_0 & 0 & 0 \\ 0 & v_0 & \varphi_0^{1/2} \\ 0 & \varphi_0^{1/2} & v_0 \end{pmatrix} w_y = 0 \tag{1.16}$$

and approximate it by

$$\tilde{w}(t+k) = \tilde{w}(t-k) - 2kQ_0\tilde{w}(t) \tag{1.17}$$

where

$$Q_0 = \begin{pmatrix} u_0 & 0 & \varphi_0^{1/2} \\ 0 & u_0 & 0 \\ \varphi_0^{1/2} & 0 & u_0 \end{pmatrix} D_{0x} + \begin{pmatrix} v_0 & 0 & 0 \\ 0 & v_0 & \varphi_0^{1/2} \\ 0 & \varphi_0^{1/2} & v_0 \end{pmatrix} D_{0y} \tag{1.18}$$

$$\equiv A\, D_{0x} + B\, D_{0y}$$

and the subscripts appended to the centered difference operator D_0 indicate the coordinate direction in which it is to operate.

We now apply Theorem 1.6 to the approximation (1.17). In this case $Q_1 \equiv 0$ so $\text{Real}(v, Q_1 v) \equiv 0$. It is easy to see that D_{0x} and D_{0y} are skew-symmetric, so $(v, Q_0 w) = -(Q_0 v, w)$ and we need only show that there is an $\eta > 0$ such that $k\|Q_0\| < 1 - \eta$ to establish stability. Using Fourier series and Parseval's relation we obtain

$$k\hat{Q}_0 = i\lambda[A \sin \xi + B \sin \eta]$$

and

$$k\|\hat{Q}_0\| = k\|Q_0\| \quad .$$

Since $\hat{Q}_0$ is normal we need only examine its eigenvalues. The eigenvalues of $k\hat{Q}_0$ are

$$\varkappa_1 = i\lambda(u_0 \sin \xi + v_0 \sin \eta)$$

$$\varkappa_{2,3} = i\lambda\left(u_0 \sin \xi + v_0 \sin \eta \pm \sqrt{\varphi_0(\sin^2\xi + \sin^2\eta)}\right)$$

so

$$k\|\hat{Q}_0\| \le \max_j |\varkappa_j| \le \lambda(|u_0| + |v_0| + \sqrt{2\varphi_0})$$

and the approximation (1.17) is stable if

$$\lambda = k/h \leq (|u_0| + |v_0| + 2\sqrt{\varphi_0})^{-1}$$

If we now replace the constants u_0, v_0 and φ_0 by smooth functions we can again establish $k\|Q_0\| < 1 - \eta$ if

$$\lambda \leq \max_{x,y,t} (|u_0(x,y,t)| + |v_0(x,y,t)| + 2\sqrt{\varphi_0(x,y,t)})^{-1}$$

and again obtain stability via Theorem 1.6.

We are close to the end of what we can establish. Consider the approximation analogous to (1.17) for the quasi-linear equation (1.15), i.e., replace u_0, v_0 and φ_0 by u, v, and φ in (1.18). If (1.15) has a smooth solution over the time interval of interest and if we substitute these values of u, v, and φ into the matrices A and B then stability follows. This is nice but life is not so easy. In the course of our computation we must use our computed approximations as coefficients and they may not be smooth enough. In practice we would invariably find that (1.17) would blow up--it would not be stable for the nonlinear problem because we cannot maintain sufficient smoothness of the approximation and hence the coefficients. The usual remedy is to modify (1.17) by adding dissipative terms of the form

$$\epsilon 2kh^2 (D_{+x}D_{-x} + D_{+y}D_{-y})\tilde{w}(x,t-k)$$

or

$$-\epsilon 2kh^2 (D_{+x}^2 D_{-x}^2 + D_{+y}^2 D_{-y}^2)\tilde{w}(x,t-k)$$

for $\epsilon > 0$ to keep the approximation smooth. In practice this usually works but there is no theoretical justification--there are no theorems which guarantee success.

2. Methods, Methods, and Methods

When we considered difference methods in the last section we simply replaced the differential operators by difference operators. From this point of view the consistency of an approximation is natural and clear. We also worked within the function space of the problem we wished to approximate. This is a computationally unrealizable convenience which made our theoretical investigations less cumbersome. However, it is a trivial matter to move into more realistic spaces and carry everything along with us. For example, let us project in the natural way from $L_2(\mathbb{R})$ into $\ell_2(Z)$, where Z is the set of integers, i.e. let $\underline{f} = \{f_j\}_{j=-\infty}^{\infty} \in \ell_2(Z)$ be defined by $f_j = f(x_j)$, $x_j = jh$, $j = 0, \pm 1, \pm 2, \ldots,$ for $f \in L_2(\mathbb{R})$. Then our L_2-estimates easily carry over to ℓ_2-estimates with $\|\underline{f}\|_{\ell_2}^2 = h \sum_j |f_j|^2$, etc.

Alternatively, considering only one space dimension for simplicity, we could derive the same methods by considering the approximations $v(x,t)$ as piecewise polynomials in x which satisfy the differential equations in the meshpoints (x_j,t), $t = 0,k,2k,\ldots$. These polynomials are "moving polynomial" interpolants which interpolate v in neighboring meshpoints. We could define $v(x,t)$ as the piecewise polynomial which, for $x_j - h/2 < x \leq x_j + h/2$, $nk - k/2 < t \leq nk + k/2$, interpolates the point values $v(x_j,t)$, $t = nk$, in chosen neighboring points. The so constructed global approximation is in general discontinuous. This is sometimes a useful way of viewing difference methods but is generally contrary to the strengths of the methods. However, there are other methods commonly used which are most naturally viewed in this way. We will now take a brief look at some such finite element and collocation methods.

The Finite Element Method

Since one of the main strengths of the finite element method is its ability to handle problems with complicated boundaries, it seems heretical to consider it in any other setting. So we consider the equation (1.1) in a bounded domain $\Omega \subset \mathbb{R}^s$ with smooth boundary and presume that boundary conditions are given which yield a well-posed problem in some Hilbert space, H_B, where the boundary conditions are satisfied. We then seek approximations in some finite dimensional subspace, S_h, of dimension N. We write our approximation in the form

$$v(x,t) = \sum_{j=1}^{N} \hat{v}_j(t)\varphi_j(x) \tag{2.1}$$

where the $\varphi_j(x)$ span S_h. In the finite element method the $\varphi_j(x)$ are piecewise polynomials with limited support and some degree of global continuity such as spline functions. They are defined over intervals in $\mathbb{R}^1$, and usually over triangles or rectangles in $\mathbb{R}^2$, etc. Having chosen S_h we only need to define a selection principle for our approximation. Here we will only consider Galerkin procedures which yield ordinary differential equations in time which can then be approximated by methods for systems of ordinary differential equations.

Instead of requiring that (2.1) satisfy the differential equation (1.1) in certain selected points the Galerkin-finite element method is based upon satisfying the weak form of the differential equation

$$(v,w_t) = (v,Pw) \tag{2.2}$$

for all $u,v \in S_h$. Sometimes it is required that (2.2) be satisfied for all v in some "test space" which might differ from S_h, but we will not consider this here. Integration by parts is used to "move" some of the derivatives from w onto v and reduce the smoothness required of the elements of S_h. One of the beauties of this approach is a direct consequence of (2.2): stability is often guaranteed, (See Strang and Fix [15], page 251.) This is in great contrast with our difficulties in establishing stability for difference methods. In the finite element approach the major theoretical difficulty is in establishing the analogue of accuracy for difference methods, which becomes the rate of convergence of S_h to H_B in this situation. That is, given $u \in H_B$ and $v \in S_h$, the rate of convergence of S_h to H_B is said to be q if

$$\|u-v\| = O(h^q) \tag{2.3}$$

where h is a measure of the discretization which can be taken as the length of the longest line contained in the support of the $\varphi_j(x)$. From the stability implied by (2.2) and the rate of convergence (2.3) of the approximating subspace, convergence results of the form of that in Theorem 1.1 follow easily, see [15].

If $V = (\hat{v}_1, \ldots, \hat{v}_n)'$ the resulting system of ordinary differential equations is of the form

$$M \frac{dV}{dt} + KV = 0 \tag{2.4}$$

where the matrices M and K are sparse but have a rather complicated structure, see [15]. In practice M and K cannot be computed exactly since they involve integrations which must be approximated. The consequences of this are discussed in Chapter 4 of [15]. Computationally the consequence of (2.4) is that we must solve systems of equations involving all our unknowns--they are "implicit". In practive this is often relaxed by various "lumping" procedures which reduce the band width of M.

A strength of the finite element method is the ease with which complicated regions Ω and nonuniform approximations can be treated. Curved elements have been developed for curved boundaries. The flexibility of the theory is wonderful in this regard. The implicit nature of (2.4) is usually unduly cumbersome for hyperbolic equations and the usual "over-stability" of the methods can pose difficulties; see page 252 of [15]. Dissipative finite element methods have been investigated by Wahlbin [19] and this is one way to get around this last difficulty--it is the approach taken with difference methods. However, finite element methods are not commonly used for hyperbolic equations at this time. The implicitness of the equations is no new problem for parabolic problems and the strong stability is often an asset. The finite element method is often used for parabolic problems for flows in porous media and engineering computations for viscous flows. The method is also used extensively in large time-dependent structures problems.

Collocation Methods

If, instead of requiring our approximation to satisfy (2.2) we ask that it satisfy the differential equation (1.1) in selected "collocation points", we are led to collocation methods. We will only consider one form of such methods here. The collocation methods most commonly used today use basis functions which are not supported locally. Complex exponentials are used for periodic boundary problems; in this case S_h is the space of trigonometric interpolants, T_N, of degree N [3]. In nonperiodic problems other choices such as Chebyshev polynomials are used [3].

The efficiency of these methods stems from the fact that trigonometric interpolants can be computed very efficiently using the well known FFT algorithm [2], the resulting approximations are smooth, and trigonometric interpolants converge rapidly for smooth functions. The accuracy of such methods follows easily from classical results but again the analysis of stability is a non trivial chore for equations with variable coefficients [6,10]. We will mention these methods again in Section 3. The use of these methods is currently being considered for numerical weather prediction, they have been used extensively for turbulence calculations and for computations of the Kortweg-de Vries equation, where some spectacular results have been obtained.

3. Methods and Algorithms

In this short section I want to mention the effect that very efficient algorithms for subproblems can have on the development of methods for the problems we are considering. One such example is the Fourier (or pseudospectral) methods mentioned in the last section. If we want to approximate the solution of the periodic boundary problem for an equation $u_t = a(x,t)u_x$ for $0 \leq x \leq 1$, $t \geq 0$, $u(0,t) = u(1,t)$ and $u(x,0) = f(x)$, we can proceed as follows. Given values of the approximation $v(x,t)$ in $2N + 1$ equally spaced points, we can interpolate $v(x,t)$ by a trigonometric interpolant of degree N , differentiate this interpolant, and then evaluate the derivative in the $2N + 1$ points. In this manner we can reduce our problem to a system of $2N + 1$ ordinary differential equations. This is the Fourier method. It is essential that one have an efficient, stable technique for computing and evaluating trigonometric interpolants if this is to be a competitive method--this service is provided by the FFT algorithm and its advent has stimulated the development and analysis of the Fourier method.

Another line of recent development is based on operator splittings which take advantage of being able to solve a reduced problem very efficiently. Operator splittings which effectively reduce problems in several space dimensions to a series of one dimensional problems have been used to advantage for some time. Such methods are discussed in the article of Lax [7] and in the books of Yanenko [21] and Marchuk [12]. Suppose we want to solve

$$u_t = (L+M)u \quad , \quad u(0) \text{ given} \tag{3.1}$$

where L and M are operators. We can write the solution in the form

$$u(t) = e^{(L+M)t} \, u(0)$$

and in particular we can write

$$u(t+k) = e^{(L+M)k} \, u(t) \, .$$

If L and M commute we can write $e^{(L+M)k} = e^{Lk} e^{Mk}$ and if we approximate e^{Lk} by $S_L(k)$ and e^{Mk} by $S_M(k)$ we are led to the approximation

$$v(t+k) = S_L(k)S_M(k)v(t) \quad .$$

This is the method of fractional steps. If each of the operators S_L and S_M is strongly stable, then so is their product. The only difficulty we have is the very strong assumption that L and M commute. However, we can look upon $e^{Lk}e^{Mk}$ as an approximation of $e^{(L+M)k}$ whether L and M commute or not, and the error committed is easily seen to be $\mathcal{O}(k^2)$ when they do not commute. Strang [14] has pointed out that $e^{Mk/2} \, e^{Lk} \, e^{Mk/2}$ is accurate through terms of order $\mathcal{O}(k^2)$ and is thus a much better form to use. Thus, we are led to the approximation

$$v(t+k) = S_M(k/2)S_L(k)S_M(k/2)v(t) \tag{3.2}$$

which is accurate of order $(q_1,2)$ if S_M and S_L are accurate of order $\mathcal{O}(h^{q_1})$. If (3.1) is a problem in two space dimensions, $L = \partial^2/\partial x^2$ and $M = \partial^2/\partial_y{}^2$ for example, then a method of the form (3.2) splits the two-dimensional problem into two one-dimensional problems. This is the most common use of splitting methods. However, it can also be used to take advantage of being able to solve a subproblem very efficiently, say $u_t = Mu$. If this is a constant coefficient problem it may be very easy to solve exactly and S_M can be chosen as e^{Mk} . One such method has been devised by Tappert [16] where he uses Fourier series to evaluate e^{Mk} -- the FFT algorithm shows up again here. Another method devised by MacCormack [9] takes advantage of the fact that a first order hyperbolic equation can easily be solved in one space dimension using the method of characteristics. MacCormack is approximating equations for viscous, compressible flow in three space dimensions. He splits the equations into a sequence of one dimensional problems and then splits each of these, taking the first order terms in M and the second order terms in L . The operator e^{Mk} is then approximated using the method of characteristics, and difference methods are used to approximate e^{Lk} . Approaches like these allow great flexibility. They enable one to take advantage of a problem's structure in an intimate manner and use special methods that only work for a very limited class of problems in more general settings.

4. Methods and Machines

Many of the time dependent problems of interest are complicated systems of equations and adequate approximations require very small "h's" or a large number of basis functions. For example, such calculations arise in aerodynamics (e.g., computing flows about airplanes), in numerical weather forecasting, and in computations of magnetohydrodynamics. The approximations for such problems give rise to very large systems of equations which must be solved to advance the solution from t to $t+k$ --there may be several million unknowns. Such calculations require nontrivial interactions between the various functional parts of a computing system -- the arithmetic and logical units, and the various elements of the hierarchy of data storage facilities ranging from the internal registers of functional units to disks or mass storage devices. This aspect of the calculations has been further complicated with the advent of machines which perform parallel operations or vector operations.

If algorithms are to be useful they must be efficiently implementable on the systems where they are to be used. The design of machines and systems has had considerable influence on the design of algorithms, and vice versa. This subject is surveyed in Budnik and Oliger [1]. Recent departures from conventional serial architectures, such as parallel and distributive computers, have called for reexamination of algorithms with regard to their overall efficiency in new computing environments. The main point I want to make here is that the traditional objectives of minimizing the number of arithmetic operations and/or the number of degrees of freedom (mesh points, etc.) to achieve given accuracy is too simplistic for the assessment of the relative merits of various methods. The data structures and interactions required for implementation must be compared and the exploitable parallelism considered.

5. Methods and Mathematics

In closing, I want to point out an area of research activity which involves detailed mathematical analysis of the solutions of approximate methods. In our discussions up to this point we have only considered approximations of smooth solutions. However, many problems of interest do not have smooth solutions -- they have contact discontinuities, shocks, etc. It is necessary to know how our methods behave in the neighborhood of such discontinuities and whether or not the resulting effects are local or global. That is, if something "funny" happens near a discontinuity, will this pollute our answers everywhere,or not? Most of the studies regarding these questions are carried out on simple model equations.

Some of this work is surveyed in Section 10 of Thomée's review article [17]. These results show that higher order accurate methods can be used to advantage for contact discontinuities and that dissipative methods are often successful in keeping the resulting computational phenomena from spreading over the domain. However, Majda and Osher [11] have recently shown that one cannot achieve more than $\mathcal{O}(h^2)$ convergence in a rarefaction region. These studies are very helpful guides for the construction of methods and provide us with insight for the largely uncharted nonlinear world. This is an area of research activity where the tools of classical and modern analysis are providing significant results, where questions are abundant and answers are rare.

References

[1] P. Budnik, Jr. and J. Oliger, "Algorithms and architecture," High Speed Computer and Algorithm Organization, D. Kuck, et al, (eds), Academic Press, New York, pp. 355-370 (1977).

[2] J. W. Cooley and J. W.Tukey, "An algorithm for the machine calculation of complex Fourier series," Math. Comp., 19, 1965, pp. 297-301.

[3] D. Gottlieb and S. A. Orszag, "Numerical analysis of spectral methods: Theory and applications," Regional Conference Series in Applied Mathematics, 24, SIAM, Philadelphia, 1977.

[4] B. Gustafsson, "The convergence rate for difference approximations to mixed initial boundary value problems," Math. Comp., 29, 1975, pp. 396-406.

[5] B. Gustafsson, H.-O. Kreiss, and A. Sundström, "Stability theory of difference approximations for mixed initial boundary value problems. II," Math. Comp., 26, 1972, pp. 649-686.

[6] H.-O. Kreiss and J. Oliger, "Stability of the Fourier method," to appear in SIAM J. Numer. Anal.

[7] P. D. Lax, "Applied Mathematics and Computing", Proc. of Symp. in Appl. Math., 20, Amer. Math. Soc., Providence, R.I., 1974, pp. 57-66.

[8] P. D. Lax and L. Nirenberg, "On stability for difference schemes: a sharp form of Garding's inequality," Comm. Pure and Appl. Math., 19, 1966, pp. 473-492.

[9] R. W. MacCormack, "An efficient numerical method for solving the time-dependent compressible Navier-Stokes equations at high Reynolds number," NASA Technical Memorandum, NASA TM X-73, 129, 1976.

[10] A. Majda, J. McDonough, and S. Osher, "The Fourier method for non-smooth initial data," to appear in Math. Comp.

[11] A. Majda and S. Osher, "Propagation of error into regions of smoothness for accurate difference approximations to hyperbolic equations," to appear.

[12] G. I. Marchuk, Methods of Numerical Mathematics, Springer-Verlag, New York, 1975.

[13] R. D. Richtmyer and K. W. Morton, Difference Methods for Initial value problems, 2nd ed., Interscience, New York, 1967.

[14] W. G. Strang, "On the construction and comparison of difference schemes," SIAM J. Numer. Anal., 5, 1968, pp. 506-517.

[15] G. Strang and G. J. Fix, An Analysis of the Finite Element Method, Prentice-Hall, Englewood Cliffs, N.J., 1973.

[16] F. Tappert, "Numerical solution of the KdV equation and its generalizations by the split-step Fourier method," Lectures in Appl. Math., 15, Amer. Math Soc., Providence, R.I., 1974, pp. 215-216.

[17] V. Thomée, "Stability theory for partial difference operators," SIAM Rev., 11, 1969, pp. 152-195.

[18] J. Varah, "Stability of difference approximations to the mixed initial boundary value problem for parabolic systems," SIAM J. Numer. Anal., 8, 1971, pp. 598-615.

[19] L. B. Wahlbin, "A dissipative Galerkin method for the numerical solution of first order hyperbolic equations," Mathematical Aspects of Finite Elements in Partial Differential Equations, C. deBoor ed., Academic Press, New York, 1974, pp. 147-170.

[20] O. Widlund, "Introduction to finite difference approximations to initial value problems for partial differential equations," Lecture Notes in Mathematics, 193, Springer-Verlag, New York, 1973.

[21] N. N. Yanenko, The Method of Fractional Steps, Springer-Verlag, New York, 1971.

Proceedings of Symposia in Applied Mathematics
Volume 22
1978

Variational Methods for Elliptic Boundary Value Problems

George J. Fix
Department of Mathematics
Carnegie-Mellon University

Outline of Lecture

1. The basic variational principles
2. Approximation
3. Rates of convergence and stability
4. Least squares approximations

1. The basic variational principles

A. Overview of lectures

Most finite element and finite difference approximations replace an elliptic boundary value problem $L_o\varphi_o = f_o$ with a finite dimensional system $L_h\varphi_h = f_h$. The discretization is normally obtained by reformulating $L_o\varphi_o = f_o$ as a variational principle for φ_o in some infinite dimensional space $\mathcal{S}$. The approximation φ_h is then obtained by selecting a finite dimensional subspace $\mathcal{S}^h$ of $\mathcal{S}$ and carrying out the variations only on $\mathcal{S}^h$. Thus the approximation involves two choices, the finite dimensional space S^h and the variational characterization of $L_o\varphi_o = f_o$. These are chosen so that the following computational objectives are satisfied:

(i) the error $\varphi_o-\varphi_h$ is "suitably small"

(ii) the finite dimensional problem $L_h\varphi_h = f_h$ is "numerically stable" and can be solved with a "reasonable amount" of computer time.

The goal of this lecture is to develop a theory which relates the practical objectives of accuracy and stability with properties of variational principles and finite element spaces S_h. The results survey the work of the author, Gunzburger, and Nicolaides [4].

B. Notation

(i) Ω is a bounded region in $\mathbb{R}^n$

(ii) $H^o(\Omega)$ = real valued functions ψ satisfying

$$\|\psi\|_o := \{\int_\Omega \psi^2\}^{1/2} < \infty$$

(iii) $\underline{H}^o(\Omega)$ = $\mathbb{R}^n$ valued functions $\underline{v}$ satisfying

$$\|\underline{v}\|_o := \{\int_\Omega \underline{v}\cdot\underline{v}\}^{1/2} < \infty$$

(iv) $H^1(\Omega)$ = real valued functions ψ satisfying

$$\|\psi\|_1 := [\int_\Omega \{\psi^2 + |\nabla\psi|^2\}]^{1/2} < \infty$$

More generally, $H^r(\Omega)$ = real valued functions ψ whose r^{th} derivatives are in $H^o(\Omega)$. We denote the norm by $\|\cdot\|_r$.

(v) If Γ denotes the boundary of Ω, then each ψ in $H^1(\Omega)$ can be defined a.e. on Γ. In fact,

$$\psi \in H^{1/2}(\Gamma)$$

(see [1]). We let $|\cdot|_{1/2}$ denote the norm on this space.

(vi) $H^{-1/2}(\Gamma)$ is the dual of $H^{1/2}(\Gamma)$ and we denote its norm by $|\cdot|_{-1/2}$.

C. Model Problem

Let Γ be divided into two parts Γ_D and Γ_N. Given

$$f_o \in H^o(\Omega),$$

find a φ_o satisfying

$$\Delta\varphi_o = f_o \quad \text{in} \quad \Omega \tag{1}$$

$$\varphi_o = 0 \quad \text{on} \quad \Gamma_D \tag{2}$$

$$\text{grad}\ (\varphi_o)\cdot\underline{\nu} = 0 \quad \text{on} \quad \Gamma_N, \tag{3}$$

where $\underline{\nu}$ is the outer normal to Γ. Alternately, find φ_o and $\underline{u}_o$ for which

$$\text{div}(\underline{u}_o) = f_o \quad \text{in} \quad \Omega \tag{4}$$

$$\text{grad}(\varphi_o) - \underline{u}_o = 0 \quad \text{in} \quad \Omega \tag{5}$$

$$\varphi_o = 0 \quad \text{on} \quad \Gamma_D \tag{6}$$

$$\underline{u}_o\cdot\nu = 0 \quad \text{on} \quad \Gamma_N \tag{7}$$

D. Dirichlet's principle

Let

$$\mathcal{S} = H^1(\Omega), \quad \mathcal{S}_o = \{\psi\in\mathcal{S} : \psi = 0 \quad \text{on} \quad \Gamma_D\} \tag{8}$$

The Dirichlet principle states that the solution φ_o minimizes

$$\frac{1}{2}\int_\Omega \text{grad}(\psi)\cdot\text{grad}(\psi) + \int_\Omega f_o\psi \tag{9}$$

for $\psi \in \mathcal{S}_o$, or what is the same, φ_o satisfies

$$\int_\Omega \text{grad}(\varphi_o)\cdot\text{grad}(\psi) + \int_\Omega f_o\psi = 0 \tag{10}$$

for all $\psi\in\mathcal{S}_o$. We rewrite (10) as

$$a(\varphi_o,\psi) + f(\psi) = 0 \tag{11}$$

E. Kelvin's Principle

This principle is in some sense dual to the Dirichlet principle with div being the "dual" of grad. Let

$$\underline{\mathcal{V}} = \underline{H}^1(\Omega)\,, \qquad \underline{\mathcal{V}}_o = \{\underline{v}\in\underline{\mathcal{V}} : \underline{v}\cdot\underline{\nu} = 0 \text{ on } \Gamma_N\} \tag{12}$$

In the Dirichlet principle we in essence had $\underline{\mathcal{V}}$ = grad $\mathcal{S}$; in the Kelvin principle the roles are reversed and we let

$$\mathcal{S}_o := \text{div}(\underline{\mathcal{V}}_o) = \{\text{div}(\underline{v}) : \underline{v}\in\underline{\mathcal{V}}_o\} \tag{13}$$

The Kelvin principle states that $\underline{u}_o :=$ grad(φ_o) minimizes

$$\int_\Omega \underline{v}\cdot\underline{v} \tag{14}$$

for $\underline{v}\in\underline{\mathcal{V}}_o$ subject to div$(\underline{v}) = f_o$. Equivalently, the pair $\{\varphi_o,\underline{u}_o\}\in\mathcal{S}_o\times\underline{\mathcal{V}}_o$ satisfies

$$\int_\Omega \{\underline{u}_o\cdot\underline{v} + \varphi_o\text{div}(\underline{v}) + \psi\text{div}(\underline{u}_o)\} = \int_\Omega \psi f_o\,. \tag{15}$$

We rewrite (15) as

$$A[(\varphi_o,\underline{u}_o)\,,\ (\psi,\underline{v})] + F(\psi,\underline{v}) = 0. \tag{16}$$

2. Approximation

A. The basic idea

For the Dirichlet principle we select a finite dimensional subspace $\mathcal{S}^h \subset \mathcal{S}$ and let

$$\mathcal{S}_o^h := \{\psi^h \in \mathcal{S}^h : \psi^h = 0 \quad \text{on} \quad \Gamma_D\} \tag{1}$$

We then seek a $\varphi_h \in \mathcal{S}_o^h$ for which

$$a(\varphi_h, \psi^h) + f(\psi^h) = 0 \quad \text{all} \quad \psi^h \in \mathcal{S}_o^h. \tag{2}$$

Once a basis $\{\psi_1^h, \ldots, \psi_N^h\}$ is chosen for $\mathcal{S}_o^h$, (2) reduces to a system in algebraic equations in N unknowns (see [2]).

Similarly, in the Kelvin principle we select a finite dimensional subspace $\underline{\mathcal{V}}^h$ of $\underline{\mathcal{V}}$ and let

$$\underline{\mathcal{V}}_o^h := \{\underline{v}^h \in \underline{\mathcal{V}}^h : \underline{v}^h \cdot \underline{\nu} = 0 \quad \text{on} \quad \Gamma_N\}, \tag{3}$$

with

$$\mathcal{S}_o^h := \operatorname{div} (\underline{\mathcal{V}}_o^h) \tag{3'}$$

We then seek a pair $\{\varphi_h, \underline{u}_h\} \in \mathcal{S}_o^h \times \underline{\mathcal{V}}_o^h$ for which

$$A((\varphi_h, \underline{u}_h), (\psi^h, \underline{v}^h)) + F(\psi^h, \underline{v}^h) = 0 \quad \text{all} \quad \{\psi^h, \underline{v}^h\} \in \mathcal{S}_o^h \times \underline{\mathcal{V}}_o^h . \tag{4}$$

Once a basis has been chosen for $\mathcal{S}_o^h \times \underline{\mathcal{V}}_o^h$ this reduces to a system of N linear equations in N unknowns, where N is the dimension of $\mathcal{S}_o^h \times \underline{\mathcal{V}}_o^h$.

B. Example

Suppose Ω is a polygon in $\mathbb{R}^2$. Subdivide Ω into triangles $T_1,\ldots,T_M$. We let $\mathcal{S}^h$ denote the finite dimensional space of continuous functions which are linear polynomials in each triangle. The dimension of this space is equal to the number k vertices in $\Omega \cup \Gamma$, and a basis $\{\psi_1^h,\ldots,\psi_k^h\}$ is obtained by letting ψ_j be the (unique) function in $\mathcal{S}^h$ which is one at the $j^{\underline{th}}$ vertex $\underline{z}_j$ and zero at all other vertices. The subspace $\mathcal{S}_o^h$ defined by (1) and needed in the Dirichlet principle is simply the functions in $\mathcal{S}^h$ which vanish at the vertices on $\overline{\Gamma}_D$. Thus $\{\psi_1^h,\ldots,\psi_N^h\}$ is a basis where $\underline{z}_1,\ldots,\underline{z}_N$ are the nodes on $\Omega \cup \Gamma_N$.

For the Kelvin principle $\underline{\mathcal{V}}^h$ could be taken as the linear space of continuous $\mathbb{R}^2$ valued functions whose components are linear polynomials in each triangle. A convenient basis for this space is the set of functions

$$\psi_j^h\binom{1}{0}, \quad \psi_j^h\binom{0}{1},$$

where ψ_j^h is defined above. A basis for the subspace $\underline{\mathcal{V}}_o^h$ defined by (3) is obtained by removing those $\underline{v}$ in the basis for $\underline{\mathcal{V}}^h$ which do not satisfy $\underline{v}\cdot\underline{\nu} = 0$ at the nodes on $\overline{\Gamma}_N$. The scalar space

$$\mathcal{S}_o^h = \operatorname{div}(\underline{\mathcal{V}}_o^h)$$

is rather interesting. It is clearly a subset of the space of discontinuous piecewise constant functions, but it may be a strict subspace of this space. See Figure 1.

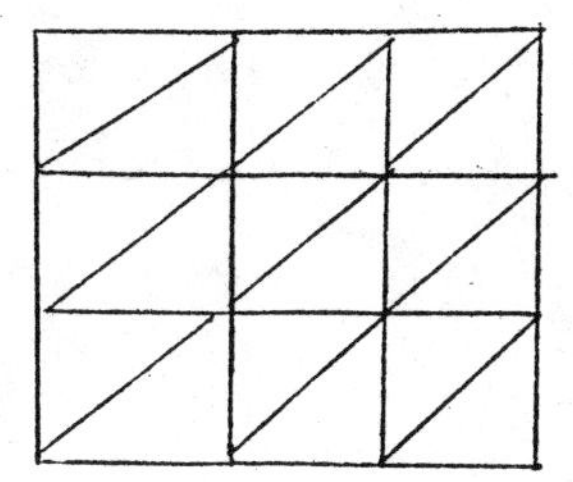

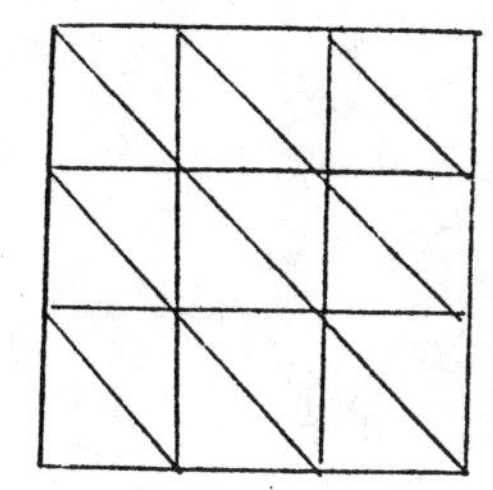

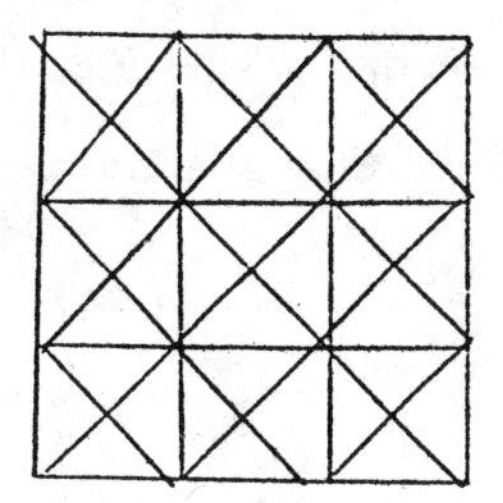

GRID 1
S_o^h = all piecewise constant functions

GRID 2
S_o^h = all piecewise constant functions

GRID 3
S_o^h = piecewise constant functions on GRIDS 1 and 2

FIGURE 1

C. Approximation properties of the subspaces

For the Dirichlet principle we shall assume there is an integer $k > 1$ and a constant $0 < C_A < \infty$ such that for any

$$\varphi \in S_o \cap H^k(\Omega), \tag{5}$$

there is a $\hat{\varphi}_h \in S_o^h$ satisfying

$$\|\varphi - \hat{\varphi}_h\|_r \leqq C_A h^{k-r} \|\varphi\|_k \qquad 0 \leq r \leq 1. \tag{6}$$

This assumption is valid with the space of piecewise linear functions defined above if $k = 2$ and if h is the largest diameter of the triangles, and in addition, if the mesh ratios are bounded (see [2],[3]).

For the Kelvin principle, we make the analogous assumption, except that the spaces are defined by (2)-(3) and (3)-(3'); i.e., for each

$$\underline{u} \in \mathcal{V}_o \cap H^k(\Omega) \tag{7}$$

we assume there is a $\hat{\underline{u}}_h \in \mathcal{V}_o^h$ for which

$$\|\underline{u}-\hat{\underline{u}}_h\|_r \leqq C_A h^{k-r} \|\underline{u}\|_k, \quad 0 \leq r \leq 1, \tag{8}$$

and for each

$$\varphi \in \mathcal{S}_o \cap H^{k-1}(\Omega) \tag{9}$$

we assume there is a $\hat{\varphi}_h \in \mathcal{S}_o^h$ for which

$$\|\varphi-\hat{\varphi}_h\|_o \leqq C_A h^{k-1} \|\varphi\|_{k-1} \tag{10}$$

These inequalities are valid for piecewise linear functions if $k = 2$. (See [2], [3], [4]) (*)

(*) We shall also assume that (6) or (8), (10) hold for integers $k^\nabla \leq k$. For example, since $k > 1$ we have in addition to (10)

$$\|\varphi-\hat{\varphi}_h\|_o \leqq C_A h \|\varphi\|_1.$$

3. Rates of convergence and stability

A. Dirichlet's principle

Virtually everything is known about this method. First the matrix that appears in the system associated with (2, Section 2) is symmetric, positive definite and hence invertible. When piecewise linear elements are used it will also be sparse, i.e., have a lot of zero entries. In addition, the method produces optimal accuracy. More precisely, we have the following theorem.

Theorem 1. There is a constant C (depending only on the constant C_A in (6)) such that

(1) $$\|\varphi_h\|_1 \leqq C \|f_o\|_{-1} \quad \text{(stability)}$$

and

(2) $$\|\varphi_o - \varphi_h\|_o \leqq Ch^k \|\varphi_o\|_k \quad \text{(accuracy in } \varphi_o)$$

(3) $$\|\text{grad}(\varphi_o) - \text{grad}(\varphi_h)\|_o \leqq Ch^{k-1} \|\varphi_o\|_k \quad \text{(accuracy in } \underline{u}_o = \text{grad}(\varphi_o) .$$

A proof of this result can be found in [2] and [3].

B. Kelvin's principle - preliminaries

The main reason for using the Kelvin principle is to achieve better accuracy in $\underline{u}_o = \text{grad}(\varphi_o)$. In many problems φ_o is a potential function and as such is not as interesting physically as its gradient, which for example could be a velocity field. The hope is that Kelvin's method will give optimal accuracy, i.e.,

(4) $$\|\underline{u}_o - \underline{u}_h\|_o \leqq Ch^k \|\underline{u}_o\|_k$$

with

$$\|\varphi_o - \varphi_h\|_o \leq Ch^{k-1} \|\varphi_o\|_{k-1}. \tag{5}$$

Observe the duality between (2)-(3) and (4)-(5). The former highlights the scalar φ_o and the latter highlights the vector $\underline{u}_o$. Unfortunately, (4)-(5) are true only in special situations.

C. Grid decomposition property

Recall that any $\underline{v} \in \mathcal{V}_o$ can be decomposed as

$$\underline{v} = \text{grad}(\xi) + \underline{z} \tag{6}$$

where

$$\text{div}(\underline{z}) = 0 \tag{7}$$

$$\int_\Omega \underline{z} \cdot \text{grad}(\xi) = 0 \tag{8}$$

Indeed, we construct ξ by solving

$$\Delta\xi = \text{div}(\underline{v}) \quad \text{in} \quad \Omega \tag{9}$$

$$\xi = 0 \quad \text{on} \quad \Gamma_D \tag{10}$$

$$\text{grad}(\xi) \cdot \underline{\nu} = 0 \quad \text{on} \quad \Gamma_N, \tag{11}$$

and letting

$$\underline{z} = \underline{v} - \text{grad}(\xi) \tag{12}$$

Observe that if

(13) $$\underline{w} = \text{grad}(\xi),$$

then from P.D.E. Theory [1]

(14) $$\|\underline{w}\|_1 \leq C\,\|\text{div}(\underline{v})\|_{-1}$$

The grid decomposition property requires that this hold on $\mathcal{V}_o^h$. More precisely, we have the following.

<u>Definition 1</u>. $\mathcal{V}_o^h$ satisfies the GDP with constant

(15) $$0 < C_G < \infty$$

if and only if for each

(16) $$\underline{v}_h \in \underline{\mathcal{V}}_o^h$$

there are

(17) $$\underline{w}_h, \underline{z}_h \in \underline{\mathcal{V}}_o^h$$

satisfying

(18) $$\underline{v}_h = \underline{w}_h + \underline{z}_h$$

with

(19) $$\text{div}(z_h) = 0, \quad \int_\Omega \underline{w}_h \cdot \underline{z}_h = 0, \quad \|\underline{w}_h\|_o \leq C_G \|\text{div}\ \underline{v}_h\|_{-1}$$

We shall show that this property is sufficient for optional convergence; (i.e., (4)-(5)) and stability. In fact, it is also necessary for the latter.

Theorem 2. Suppose $\underline{\mathcal{V}}_o^h$ satisfies the GDP with constant C_G. Then the Kelvin approximation is stable in the following sense. For each

$$f_h \in \mathcal{S}_o^h := \mathrm{div}(\mathcal{V}_o^h) \tag{20}$$

let $\underline{v}_h \in \underline{\mathcal{V}}_o^h$ minimize $\|\underline{v}_h\|_o$ subject to $\mathrm{div}(\underline{v}_h) = f_h$. Then

$$\|\underline{v}_h\|_o \leqq C_G \|f_h\|_{-1} \tag{21}$$

Conversely, if the Kelvin principle is stable with constant C_G, then the GDP holds with constant C_G.

For linear elements the GDP may or may not hold. For example, it does not hold for Grids 1 and 2 in (Figure 1, Section 2). It does, however, hold for Grid 3. (See [4]).

Theorem 3. Let $\underline{\mathcal{V}}_o^h$, $\mathcal{S}_o^h = \mathrm{div}(\underline{\mathcal{V}}_o^h)$ satisfy the approximation properties [(7)-(10), Section 2]. Then

$$\|\mathrm{div}(\underline{u}_o - \underline{u}_h)\|_{-1} \leqq Ch^k \|\underline{u}_o\|_k \tag{22}$$

If in addition the GDP holds, then

$$\|\underline{u}_o - \underline{u}_h\|_o \leqq Ch^k \|\underline{u}_o\|_k \tag{23}$$

$$\|\varphi_o - \varphi_h\|_o \leqq Ch^{k-1} (\|\underline{u}_o\|_k + \|\varphi_o\|_{k-1}) \tag{24}$$

4. Least squares approximations

This method is in essense a mixture of the Dirichlet and Kelvin principles and it shares characteristics of both. In its simplest form we have

(1) $$\mathcal{S}_o^h = \{\psi^h \in \mathcal{S}^h : \psi^h = 0 \text{ on } \Gamma_D\}, \quad \underline{\mathcal{V}}_o^h = \{\underline{v}^h \in \underline{\mathcal{V}}^h : \underline{v}^h \cdot \underline{\nu} = 0 \text{ on } \Gamma_N\}$$

and we seek the minimum of

(2) $$\int_\Omega \{|\operatorname{grad}\psi^h - \underline{v}^h|^2 + |\operatorname{div} \underline{v}^h - f_o|^2\}$$

over $\{\psi^h, \underline{v}^h\} \in \mathcal{S}_o^h \times \underline{\mathcal{V}}_o^h$. Letting $\{\varphi_h, \underline{u}_h\}$ denote the pair where the minimum is achieved we have

(3) $$B(\{\varphi_h, \underline{u}_h\}, \{\psi^h, \underline{v}^h\}) = \int_\Omega f \operatorname{div} \underline{v}^h \quad \text{all } \{\psi^h, \underline{v}^h\} \in \mathcal{S}_o^h \times \underline{\mathcal{V}}_o^h$$

where

(4) $$B(\{\varphi_h, \underline{u}_h\}, \{\psi^h, \underline{v}^h\}) := \int_\Omega \{(\operatorname{grad} \varphi_h - u_h) \cdot (\operatorname{grad} \psi^h - \underline{v}^h) + \operatorname{div} \underline{u}_h \cdot \operatorname{div} v_h\}$$

As is typical of least square methods we get a best approximation in a rather crazy norm, i.e.,

(5) $$|||(\psi^h, \underline{v}^h)||| := B(\{\psi^h, \underline{v}^h\}, \{\psi^h, \underline{v}^h\})^{1/2}$$

More precisely, we have the following theorem.

Theorem 4. Let $\{\varphi_h,\underline{u}_h\}$ satisfy (3). Then

$$|||\{\varphi_o-\varphi_h,\underline{u}_o-\underline{u}_h\}||| \leq |||\{\varphi_o-\psi^h,\underline{u}_o-\underline{v}^h\}||| \tag{6}$$

for all $\{\psi^h,\underline{v}^h\}\in S_o^h \times \underline{U}_o^h$.

For simplicity let us assume that [(6), Section 2], i.e.,

$$\|\varphi_o-\hat{\varphi}_k\|_r \leq C_A h^{k-r} \|\varphi_o\|_k \text{ , } 0 \leq r \leq 1 \text{ ,} \tag{7}$$

is valid for some $\hat{\varphi}_h\in S_o^h$, and that

$$\|\underline{u}_o-\hat{\underline{u}}_h\|_r \leq C_A h^{k-r}\|\underline{u}_o\|_k \quad 0 \leq r \leq 1 \tag{8}$$

is also valid for some $\hat{\underline{u}}_h\in\underline{U}_o^h$.

Corollary

$$\{\varphi_o-\varphi_h,\underline{u}_o-\underline{u}_h\} \leq C_A h^{k-1}\{\|\varphi_o\|_k+\|\underline{u}_o\|_k\} \tag{9}$$

Remarkably the least square method will always give optimal approximations to φ_o -- this it inherits from the Dirichlet principle -- but it requires the G.D.P. to get optimal accuracy in $\underline{u}_o$ -- this it inherits from the Kelvin principle.

Theorem 5. Let $\underline{U}_o^h$, S_o^h satisfy (7)-(8). Then

$$\|\varphi_o-\varphi_h\|_o \leq Ch^k(\|\varphi_o\|+\|\underline{u}_o\|_k) \tag{10}$$

$$\|\mathrm{div}(\underline{u}_o-\underline{u}_h)\|_{-1} \leq Ch^k\|\underline{u}_o\|_k \tag{11}$$

If in addition, the GDP holds

$$\|\underline{u}_o-\underline{u}_h\|_{-1} \leq Ch^k\|\underline{u}_o\|_k. \tag{12}$$

References

[1] Lions, J.L. and E. Nagenes (1968). Problems aux limites non homogenes et applications, Durod, Paris.

[2] Strang, G. and G. Fix (1973). An Analysis of the Finite Element Method, Wiley, Englewood Cliffs.

[3] The Mathematical Foundations of the Finite Element Method (University of Maryland at Baltimore), Academic Press, New York, 1973.

[4] Fix, G., M. Gunzburger, and R.A. Nicolaides, "On finite element approximations to first order systems of the elliptic type"
I. Least Squares Methods
II. Galerkin Methods
NASA-ICASE Research Reports, 1977.

[5] Jespersen, D., "On least square decomposition of elliptic boundary value problems," to appear Math. of Comp.

[6] Brezzi, F., "On the existence, uniqueness and approximation of saddle point problems", R.A.I.R.O., 1974, 129-151.

Notes

[1] Jespersen [4] first showed that the least squares method gave optimal accuracy in φ_o; this work was extended in [4] to include more general equations and estimates for $\underline{u}_o$. This work is the basis for results presented in Section 4.

[2] Brezzi [6] has sufficient conditions for the convergence of $\underline{u}_h$ to $\underline{u}_o$. The order of accuracy, however, is only $O(h)$ in L_2 for linear elements. This appears to be sharp in general, however, with the G.D.P. $O(h^2)$ is obtained as is indicated in Section 3. The latter is based on the work in [4].

Appendix I

Proof of Theorem 2

(1) Let $\mathcal{N}_h(\text{div}) = \{\underline{v}_h \in \underline{\mathcal{V}}_o^h : \text{div}(\underline{v}_h) = 0 \text{ in } \Omega\}$

and let $\mathcal{N}_h(\text{div})^\perp$ denote its orthogonal complement so

(2) $$\underline{\mathcal{V}}_o^h = \mathcal{N}_h(\text{div}) \oplus \mathcal{N}_h(\text{div})^\perp$$

First suppose the GDP holds, i.e., any $\underline{v}_h \in \underline{\mathcal{V}}_o^h$ can be written

(3) $$\underline{v}_h = \underline{w}_h + \underline{z}_h,$$

where

(4) $$\underline{z}_h \in \mathcal{N}_h(\text{div}), \int_\Omega \underline{w}_h \cdot \underline{z}_h = 0, \quad \|\underline{w}_h\|_o \leq c_G \|\text{div } \underline{v}_h\|_{-1}.$$

Moreover, let $\underline{v}_h \in \underline{\mathcal{V}}^h$ satisfy

(5) $$\|\underline{v}_h\|_o = \text{min. subject to } \underline{v}_h \in \mathcal{V}_o^h \text{ and } \text{div}(\underline{v}_h) = f_h,$$

where $f_h \in \mathcal{S}_o^h = \text{div}(\underline{\mathcal{V}}^h)$ is given. We want to show that

(6) $$\|\underline{v}_h\|_o \leq c_G \|f_h\|_{-1}.$$

To do this we write $\underline{v}_h$ as in (3)-(4). The claim is that $\underline{z}_h = \underline{0}$ and so

(7) $$\|\underline{v}_h\|_o = \|\underline{w}_h\|_o \leq c_G \|\text{div } \underline{v}_h\|_{-1} = c_G \|f_h\|_{-1}$$

To see this observe that for any real number δ, $\underline{v}_h+\delta\underline{z}_h$ is in $\underline{\mathcal{V}}_o^h$ and

(8) $$\operatorname{div}(\underline{v}_h+\delta\underline{z}_h) = \operatorname{div}(\underline{v}_h) = f_h$$

Thus

(9) $$\int_\Omega (\underline{v}_h+\delta\underline{z}_h)\cdot(\underline{v}_h+\delta\underline{z}_h) \geq \int_\Omega \underline{v}_h\cdot\underline{v}_h,$$

i.e.,

(10) $$2\delta\int_\Omega \underline{z}_h\cdot\underline{v}_h \geq -\delta^2\int_\Omega \underline{z}_h\cdot\underline{z}_h$$

Since δ is arbitrary we necessarily have

(11) $$\int_\Omega \underline{z}_h\cdot\underline{v}_h = 0$$

But $\underline{v}_h = \underline{w}_h + \underline{z}_h$ and $\underline{w}_h$ is orthogonal to $\underline{z}_h$. This means

(12) $$\int_\Omega \underline{z}_h\cdot\underline{z}_h = \int_\Omega (\underline{v}_h-\underline{w}_h)\cdot\underline{z}_h = 0$$

Conversely, assume that the Kelvin problem is stable (with constant C_G) and let $\underline{v}_h\in\underline{\mathcal{V}}_o^h$ be given. By (2), we can always write

(13) $$\underline{v}_h = \underline{w}_h + \underline{z}_h,$$

where

(14) $$\underline{w}_h\in\eta_h(\operatorname{div})^\perp, \quad \underline{z}_h\in\eta_h(\operatorname{div}).$$

We want to select $\underline{w}_h$ such that

$$\|w_h\|_o \leqq c_G \|\text{div } \underline{v}_h\|_{-1}$$

To do this we solve a Kelvin problem. More precisely, let

$$f := \text{div}(\underline{v}_h) ,$$

and let $\underline{w}_h$ minimize $\|\underline{w}_h\|_o$ subject to

$$\underline{w}_h \in \underline{\mathcal{V}}_o^h, \quad \text{div}(\underline{w}_h) = f_h.$$

By (6) ($\underline{w}_h$ is playing the role of $\underline{v}_h$ in this inequality)

$$\|w_h\|_o \leqq c_G \|f_h\|_{-1}$$

Moreover,

$$\underline{z}_h := \underline{v}_h - \underline{w}_h \in \mathcal{N}_h(\text{div})$$

Appendix II

Proof of Theorem 3

The key identity that will be used repeatedly is

$$(1)\quad \int_\Omega \{\underline{u}_o\cdot\underline{v}^h+\varphi_o \operatorname{div}(\underline{v}^h)+\psi^h \operatorname{div}\underline{u}_o\} = \int_\Omega \{\underline{u}_h\cdot\underline{v}^h + \varphi_h \operatorname{div}\underline{v}^h + \psi^h \operatorname{div}\underline{u}_h\}.$$

This is valid for all $\{\psi^h,\underline{v}^h\}\in S_o^h \times \underline{V}_o^h$ (since both sides are equal to $\int_\Omega f_o\psi^h$ by [(15), Section 1] and [(4), Section 2]).

Lemma 1. For all $\underline{w}^h\in\underline{V}_o^h$

$$(2)\quad \|\operatorname{div}(\underline{u}_o-\underline{u}_h)\|_o \leqq \|\operatorname{div}(\underline{u}_o-\underline{w}^h)\|_o$$

In particular,

$$(3)\quad \|\operatorname{div}(\underline{u}_o-\underline{u}_h)\|_o \leqq \|\operatorname{div}(\underline{u}_o-\hat{\underline{u}}_h)\|_o ,$$

where $\hat{\underline{u}}_h$ is the function in [(8), Section 2].

Proof. Let $\underline{v}^h = 0$ in (1). Then

$$(4)\quad \int_\Omega \operatorname{div}(\underline{u}_o-\underline{u}_h)\,\psi^h = 0$$

all $\psi^h\in S^h = \operatorname{div}(\underline{V}_o^h)$. Let $\psi^h = \underline{u}_h-\underline{w}_h$. Then (4) gives (2).

Lemma 2

$$(5)\quad \|\operatorname{div}(\underline{u}_o-\underline{u}_h)\|_{-1} \leqq C_A h\|\operatorname{div}(\underline{u}_o-\underline{u}_h)\|_o$$

Proof

Solve
$$-\Delta\xi + \xi = \operatorname{div}(\underline{u}_o-\underline{u}_h) \quad \text{in } \Omega$$
$$\xi = 0 \quad \text{on } \Gamma$$

Then

$$\|\xi\|_1 = \|\mathrm{div}(\underline{u}_o - \underline{u}_h)\|_{-1} \tag{6}$$

But

$$\|\xi\|_1^2 = \int_\Omega \{\nabla\xi\cdot\nabla\xi + \xi^2\} = \int_\Omega \xi \,\mathrm{div}(\underline{u}_o - \underline{u}_h) \tag{7}$$

We note that if $\underline{v}^h = \underline{0}$ in (1)

$$\int_\Omega \psi^h \,\mathrm{div}(\underline{u}_o - \underline{u}_h) = 0 \quad \text{all} \quad \psi^h \in S_o^h \tag{8}$$

Thus letting $\psi^h = \hat{\xi}^h$

$$\|\xi\|_1^2 = \int_\Omega (\xi - \hat{\xi}^h)\,\mathrm{div}(\underline{u}_o - \underline{u}_h) \tag{9}$$

$$\leq \|\xi - \hat{\xi}^h\|_o \,\|\mathrm{div}(\underline{u}_o - \underline{u}_h)\|_o$$

Using the approximation property (10) with $k = 1$ we can choose $\hat{\xi}^h$ such that

$$\|\xi - \hat{\xi}^h\|_o \leq C_A h \|\xi\|_1 \tag{10}$$

Thus (5) follows from (6), (9), and (10).

Observe that combining Lemmas 1 and 2 we have the first part of Theorem 3; i.e.,

$$(11)\qquad \begin{aligned} \|\operatorname{div}(\underline{u}_o-\underline{u}_h)\|_{-1} &\leqq C_A h\|\operatorname{div}(\underline{u}_o-\underline{u}_h)\|_o && \text{(Lemma 2)}\\ &\leqq C_A h\|\operatorname{div}(\underline{u}_o-\hat{\underline{u}}_h)\|_o && \text{(Lemma 1)}\\ &\leqq C_A h\|\underline{u}_o-\hat{\underline{u}}_h\|_1 \\ &\leqq C_A^2 h^k\|\underline{u}_o\|_k && \text{([8], Section 2)} \end{aligned}$$

So far the GDP has not been used, however from this point on it will play a crucial role. In particular, write

$$(12)\qquad \underline{u}_h-\hat{\underline{u}}_h = \underline{w}_h + \underline{z}_h$$

where

$$(13)\qquad \operatorname{div}(\underline{z}_h) = 0,\ \int_\Omega \underline{w}_h\cdot\underline{z}_h = 0,\ \|\underline{w}_h\|_o \leqq C_G\|\operatorname{div}(\underline{u}_h-\hat{\underline{u}}_h)\|_{-1}$$

Observe that

$$(14)\qquad \begin{aligned} \|w_h\|_o &\leqq C_G\|\operatorname{div}(\underline{u}_h-\hat{\underline{u}}_h)\|_{-1}\\ &\leqq C_G\{\|\operatorname{div}(\underline{u}_h-\underline{u}_o)\|_{-1} + \|\operatorname{div}(\underline{u}_o-\hat{\underline{u}}_h)\|_{-1}\}\\ &\leq C_G\{\|\operatorname{div}(\underline{u}_h-\underline{u}_o)\|_{-1} + \|\underline{u}_o-\hat{\underline{u}}_h\|_o\}\\ &\leq Ch^k\|\underline{u}_o\|_k \end{aligned}$$

Thus it is sufficient to obtain a similar bound for $\underline{z}_h$. In particular, letting $\psi^h = 0$ and $\underline{v}^h = \underline{z}_h$ in (1) we obtain

$$(15)\qquad \int_\Omega \underline{u}_h\cdot\underline{z}_h = \int_\Omega \underline{u}_o\cdot\underline{z}_h$$

Thus

$$\int_\Omega \underline{z}_h \cdot \underline{z}_h = \int_\Omega (\underline{u}_h - \hat{\underline{u}}_h) \cdot z_h = \int_\Omega (\underline{u}_o - \hat{\underline{u}}_h) \cdot \underline{z}_h \tag{16}$$

This gives

$$\|\underline{z}_h\|_o \leqq \|\underline{u}_o - \hat{\underline{u}}_h\|_o \tag{17}$$

Inequalities (12)-(14) and (17) give

$$\|\underline{u}_h - \hat{\underline{u}}_h\|_o \leqq Ch^k \|\underline{u}_o\|_k \tag{18}$$

Hence [(23), Section 3] follows from the triangle inequality

To estimate $\varphi_o - \varphi_h$ we let $\psi^h = 0$ in (1) to get

$$\int_\Omega \{\varphi_h \operatorname{div} \underline{v}^h\} = \int_\Omega \{\varphi_o \operatorname{div} \underline{v}^h + \underline{v}^h \cdot (\underline{u}_o - \underline{u}_h)\} \tag{19}$$

Let $\hat{\varphi}_h$ be the function defined in [(10), Section 2] with $\varphi = \varphi_o$. Then

$$\int_\Omega \{(\varphi_h - \hat{\varphi}_h) \operatorname{div} \underline{v}^h\} = \int_\Omega \{(\varphi_o - \hat{\varphi}_h) \operatorname{div}(\underline{v}^h) + \underline{v}^h \cdot (\underline{u}_o - \underline{u}_h)\} \tag{20}$$

Since $\mathcal{S}_o^h = \operatorname{div}(\mathcal{V}_o^h)$ there is a $\underline{v}_h \in \mathcal{V}_o^h$ such that

$$\varphi_h - \hat{\varphi}_h = \operatorname{div}(\underline{v}_h) \tag{21}$$

We now use the GDP to write

$$\underline{v}_h = \underline{w}_h + \underline{z}_h, \tag{22}$$

with

$$\operatorname{div}(\underline{z}_h) = 0, \quad \int_\Omega \underline{w}_h \cdot \underline{z}_h = 0, \quad \|\underline{w}_h\|_o \leqq c_G \|\varphi_h - \hat{\varphi}_h\|_{-1} \tag{23}$$

Letting $\underline{v}^h = \underline{w}_h$ in (20) we obtain

$$\begin{aligned} \int_\Omega |\varphi_h - \hat{\varphi}_h|^2 &\leqq \|\varphi_o - \hat{\varphi}_h\|_o \, \|\varphi_h - \hat{\varphi}_h\|_o + \|\underline{w}_h\|_o \|\underline{u}_o - \underline{u}_h\|_o \\ &\leqq \|\varphi_o - \hat{\varphi}_h\|_o \, \|\varphi_h - \hat{\varphi}_h\|_o + c_G \|\varphi_h - \hat{\varphi}_h\| \, \|\underline{u}_o - \underline{u}_h\|_o \end{aligned} \tag{24}$$

Thus

$$\|\varphi_h - \hat{\varphi}_h\|_o \leqq \|\varphi_o - \hat{\varphi}_h\|_o + c_G \|\underline{u}_o - \underline{u}_h\|_o \tag{25}$$

The triangle inequality now establishes [(24, Section 3].

Appendix III

Proof of Theorem 5

The proof is remarkably similar to the proof of Theorem 3. In particular, we will be able to estimate $\text{div}(\underline{u}_o-\underline{u}_h)$ in a negative norm without the GDP. The latter is then used to get an L_2 estimate for $\underline{u}_o-\underline{u}_h$.

The key identity is

$$(1) \quad B((\varphi_h,\underline{u}_h),\ (\psi^h,\underline{v}^h)) = B((\varphi_o,\underline{u}_o),\ (\psi^h,\underline{v}^h)) \text{ all } (\psi^h,\underline{v}^h)\in S_o^h\times \underline{V}_o^h,$$

where

$$(2) \quad B((\varphi,\underline{u}),\ (\psi,\underline{v})) := \int_\Omega \{[\text{grad } \varphi-\underline{u}]\cdot[\text{grad } \psi-\underline{v}] + \text{div } \underline{u}\cdot \text{div } \underline{v}\}$$

Letting

$$(3) \qquad \epsilon := \varphi_o-\varphi_n, \qquad \underline{e} := \underline{u}_o-\underline{u}_h,$$

(1) is equivalent to

$$(4) \qquad B((\epsilon,\underline{e}),\ (\psi^h,\underline{v}^h)) = 0 \text{ all } (\psi^h,\underline{v}^h)\in S_o^h \times \underline{V}_o^h$$

We know that

$$(5) \qquad |||(\epsilon,\underline{e})||| \leq Ch^{k-1}(\|\underline{u}_o\|_k + \|\varphi_o\|_k),$$

hence it is sufficient to bound errors in terms of $|||(\epsilon,e)|||$.

Lemma 1

$$\|\operatorname{div} \underline{e}\|_{-1} \leq Ch\, |||(\epsilon, e)||| \tag{6}$$

Proof. Let

$$\xi \in H^1(\Omega) \tag{7}$$

be given. Solve

$$\Delta\alpha = \xi \quad \text{in} \quad \Omega \tag{8}$$

$$\alpha = 0 \quad \text{on} \quad \Gamma_D \tag{9}$$

$$\underline{a}\cdot\underline{\nu} = 0 \quad \text{on} \quad \Gamma_N, \tag{10}$$

where $\underline{a} = \operatorname{grad} \alpha$. Note that $\{\alpha, \underline{a}\} \in S_o \times \underline{V}_o$ and

$$B((\epsilon, \underline{e}), (\alpha, \underline{a})) = \int_\Omega \operatorname{div} \underline{a} \operatorname{div} \underline{e} = \int_\Omega \xi \operatorname{div} \underline{e} \tag{11}$$

Using (4) we have

$$\begin{aligned} B((\epsilon, \underline{e}), (\alpha, \underline{a}) &= B((\epsilon, \underline{e}), (\alpha - \hat{\alpha}_h, \underline{a} - \hat{\underline{a}})) \\ &\leq |||(\epsilon, \underline{e})|||\ \ |||(\alpha - \hat{\alpha}_h, \underline{a} - \hat{\underline{a}}_h)||| \end{aligned} \tag{12}$$

We choose $(\hat{\alpha}_h, \hat{\underline{a}}_h) \in S_o^h \times \underline{V}_o^h$ so that

$$\begin{aligned} |||(\alpha - \hat{\alpha}_h, \underline{a} - \hat{\underline{a}}_h||| &\leq Ch(\|\alpha\|_2 + \|\underline{a}\|_2) \\ &\leq Ch\, \|\alpha\|_3 \end{aligned} \tag{13}$$

From PDE theory [1] we have

$$\|\alpha\|_3 \leq C\|\xi\|_1 \tag{14}$$

provided Ω is sufficiently smooth. Thus

$$|\int_\Omega \xi \operatorname{div} \underline{e}| \leq Ch \|\xi\|_1 \; |||(\epsilon,\underline{e})||| \tag{15}$$

and so

$$\|\operatorname{div} \underline{e}\|_{-1} = \sup_{\|\xi\|_1 \leq 1} |\int_\Omega \xi \operatorname{div} \underline{e}| \leq Ch|||(\epsilon,\underline{e})||| \tag{16}$$

Lemma 2

$$\|\epsilon\|_o \leq Ch|||(\epsilon,\underline{e})||| \tag{17}$$

Proof. Solve

$$\Delta\xi = \epsilon \quad \text{in} \quad \Omega \tag{18}$$

$$\xi = 0 \quad \text{on} \quad \Gamma \tag{19}$$

Then

$$B((\epsilon,\underline{e}), (\xi,\underline{0})) = \int_\Omega (\operatorname{grad} \epsilon - \underline{e}) \operatorname{grad} \xi \tag{20}$$

But from (18)-(19),

$$\int_\Omega \operatorname{grad} \epsilon \cdot \operatorname{grad} \xi = - \int_\Omega \epsilon^2 \tag{21}$$

On the other hand,

$$\int_\Omega \underline{e} \cdot \operatorname{grad} \xi = - \int_\Omega \operatorname{div} (\underline{e}) \xi \tag{22}$$

$$\leq \|\operatorname{div} \underline{e}\|_{-1} \|\xi\|_1$$

Finally,

$$(23)\qquad \begin{aligned} B((\epsilon,\underline{e}),(\xi,\underline{0})) &= B((\epsilon,\underline{e}),(\xi-\hat{\xi}_h,\underline{0}))\\ &\leq |||\epsilon,\underline{e}|||\ |||(\xi-\hat{\xi}_h,0)|||\\ &\leq Ch|||(\epsilon,\underline{e})|||\ \|\xi\|_2\\ &\leq Ch|||(\epsilon,\underline{e})|||\ \|\epsilon\|_o \end{aligned}$$

Combining (21)-(23) we obtain (17).

Lemma 3. Let the GDP hold. Then

$$(17)\qquad \|\underline{e}\|_o \leq Ch^k\|\underline{u}_o\|_k$$

Proof. Write

$$\underline{u}_h-\hat{\underline{u}}_h = \underline{w}_h + \underline{z}_h,$$

where

$$\int_\Omega \underline{w}_h\cdot\underline{z}_h = 0,\ \mathrm{div}(\underline{z}_h) = 0,\quad \|w_h\|_o \leq C_G\ \|\mathrm{div}(\underline{u}_h-\hat{\underline{u}}_h)\|_{-1}.$$

Thus

$$\begin{aligned} \|\underline{w}_h\|_o &\leq C_G(\|\mathrm{div}(\underline{u}_o-\hat{\underline{u}}_h)\|_{-1} + \|\mathrm{div}\ \underline{e}\|_{-1})\\ &\leq C_G(C_A h^k\ \|\underline{u}_o\|_k + Ch|||(\epsilon,\underline{e})|||). \end{aligned}$$

To estimate $\|\underline{z}_h\|_o$ we let $\underline{v}^h = \underline{z}_h$ and $\psi^h = 0$ in (1). This gives (since $\mathrm{div}\ \underline{z}_h = 0$)

$$\int_\Omega \underline{u}_h\cdot\underline{z}_h = \int_\Omega \underline{u}_o\cdot\underline{z}_h$$

Thus

$$\|z_h\|_o \leq \|\underline{u}_o - \hat{\underline{u}}_h\|_o$$

(as in Appendix II).

BCDEFGHIJ-CM-89876543210